EGE NO

Site Management of Building Services Contractors

JOIN US ON THE INTERNET VIA WWW, GOPHER, FTP OR EMAIL:

WWW: http://www.thomson.com
GOPHER: gopher.thomson.com
FTP: ftp.thomson.com
EMAIL: findit@kiosk.thomson.com

A service of I(T)P®

Site Management of Building Services Contractors

Jim Wild
Eur Ing, CEng, FCIBSE, MIMgt

E & FN SPON
An Imprint of Chapman & Hall
London · Weinheim · New York · Tokyo · Melbourne · Madras

Published by E & FN Spon, an imprint of Chapman & Hall, 2–6 Boundary Row, London SE1 8HN, UK

Chapman & Hall, 2–6 Boundary Row, London SE1 8HN, UK

Chapman & Hall GmbH, Pappelallee 3, 69469 Weinheim, Germany

Chapman & Hall USA, 115 Fifth Avenue, New York, NY 10003, USA

Chapman & Hall Japan, ITP-Japan, Kyowa Building, 3F, 2-2-1 Hirakawacho, Chiyoda-ku, Tokyo 102, Japan

Chapman & Hall Australia, 102 Dodds Street, South Melbourne, Victoria 3205, Australia

Chapman & Hall India, R. Seshadri, 32 Second Main Road, CIT East, Madras 600 035, India

First edition 1997

© 1997 L.J. Wild
HMSO (The Essentials of Health and Safety at Work)

Typeset in 10½/12pt Sabon by Acorn Bookwork, Salisbury, Wiltshire
Printed in Great Britain by Alden Press, Oxford

ISBN 0 419 20450 4

Apart from any fair dealing for the purposes of research or private study, or criticism or review, as permitted under the UK Copyright Designs and Patents Act, 1988, this publication may not be reproduced, stored, or transmitted, in any form or by any means, without the prior permission in writing of the publishers, or in the case of reprographic reproduction only in accordance with the terms of the licences issued by the Copyright Licensing Agency in the UK, or in accordance with the terms of licences issued by the appropriate Reproduction Rights Organization outside the UK. Enquiries concerning reproduction outside the terms stated here should be sent to the publishers at the London address printed on this page.

The publisher makes no representation, express or implied, with regard to the accuracy of the information contained in this book and cannot accept any legal responsibility or liability for any errors or omissions that may be made.

A catalogue record for this book is available from the British Library

Library of Congress Catalog Card Number: 96-68678

Printed on permanent acid-free text paper, manufactured in accordance with ANSI/NISO Z39. 48-1992 and ANSI/NISO Z39.48-1984 (Permanence of Paper).

BASFORD HALL COLLEGE
LIBRARY

ACC. No.

CLASS No. 690 06 WIL

Contents

List of figures

List of tables

List of abbreviations

ACE	Association of Consulting Engineers
ACoP	Approved code of practice
AGV	Automatic Guided Vehicle
AHU	Air handling unit
AI	Architect's instruction
BCO	Building Control Officer
B of Q	Bill of quantities
BRE	The Building Research Establishment
BREEAM	Building Research Establishment Environmental Assessment Method
BRECSU	Building Research Establishment Conservation Support Unit
BS	British Standards (when followed by a number)
BS	Building services
BSM	Building services manager
BMS	Building management system
BSRIA	The Building Services Research and Information Association
BWIC	Builders' work in connection with building services
CA	Contract administrator
CAD	Computer aided design
CAWS	Common arrangement work sections
CCPI	Co-ordinating Committee for Project Information
CCTV	Closed circuit television
CDM	Construction (Design & Management) Regulations, 1994
CIBSE	The Chartered Institution of Building Services Engineers
CIRIA	The Construction Industry Research and Information Association
CM	Construction management Commissioning management (in Chapter 9)
CORGI	Confederation of Registered Gas Installers
COPs	Codes of practice
CPD	Continuing professional development
COSHH	Control of substances hazardous to health
CW	Cold water

D & B	Design & build
DLP	Defects liability period
DIY	Do it yourself
DT	Design team
DW	Drinking water
E	Electrical (services)
ECA	Electrical Contractors Association
EEO	The Energy Efficiency Office
EHO	Environmental Health Officer
EI	Engineer's instruction
EP	Environmental Plan
ER	Employer's representative
F	Fire (services)
FCO	Fire Control Officer
FCU	Fan coil units
FRS	The Fire Research Station
GA	General arrangement (drawing)
GF	General foreman
GMP	Guaranteed maximum price
HCFC	Hydrochlorofluorocarbons
HFA	Hand fire appliances
HFC	hydrofluorocarbon
HS	Health & safety
HSE	Health and Safety Executive
HV	High voltage
HSW	Health & Safety at Work Act etc., 1974
HWS	Hot water service
HVAC	Heating, ventilation and air conditioning
HVCA	Heating & Ventilating Contractors Association
IEE	Institution of Electrical Engineers
IR	Infrared
L	Lifts (vertical transportation services, including escalators)
LPC	Loss Prevention Council
LV	Low voltage
M	Mechanical (services)
M & E	Mechanical and electrical
MC	Management contract
MCC	Motor control centre
MTL	Manufacturer's technical literature
NALM	National Association of Lift Makers
P	Public health (services, including sanitation and water)
PA	Public address
PABX	Private automatic branch exchange

PC	Principal contractor
PCI	Principal contractor's instruction
PM	Project manager
PPE	Personal protective equipment
PREFAB	Prefabricated work
PS	Planning supervisor
PSA	Property Services Agency
QP	Quality plan
QS	Quantity surveyor
RAD	Radiator
RC	Reinforced concrete
RFI	Request for information
RIBA	Royal Institute of British Architects
SC	Subcontractor
SI	Standing instructions
SP	Safety plan
SW & V	Soil, waste and vent
TA	Technical author
TM	Technical Memorandum
TN	Technical Note
TQS	Technical query sheet
UPS	Uninterrupted power system
VAV	Variable air volume
VCD	Volume control damper
VRF	Variable refrigerant flow
VRV	Variable refrigerant volume

Preface

Some good initiatives have come about in the last decade and are well placed to make valued contributions to the management of the design and construction of mechanical and electrical services in the post-Latham era. One example is guidance such as the Building Services Research and Information Association's (BSRIA) *The Allocation of Design Responsibilities* TN 8/94 – *a code of conduct to avoid conflict* – with its objective of sharpening the fuzzy edges between the design responsibilities of designers and installers. The Chartered Institute of Building (CIOB) *Code of Practice for Project Management* has been well received. In recognition of clients and the need to be politically correct, the Association of Consulting Engineers (ACE) has revised and opened up its fee scales to enable its members to provide a more bespoke service. Not only have these leading research and professional bodies made worthwhile contributions, but contracting and manufacturing organizations such as the Electrical Contractors Association (ECA), with its *Lean Construction* Report, and the National Association of Lift Makers (NALM), with the launch of new guidance documents in 1994, have been major playmakers.

The embracing of quality management systems has brought some improvement to the industry and although it still has far to go now travels in company with the Construction Design and Management Regulations 1994 (CDM) and the uptake of Environmental Management Systems BS7750. These will have a great impact on building design, construction, operation and maintenance.

By virtue of their uniqueness, in that every one is a prototype, the constructing of buildings will never hold the promise of an easy vocation for building site managers. They will continue to strive to deliver the specified products on time and to cost against a background of increasing legislation, and under the critical stare of the client and the professional team. Nowhere is the critical stare longer or harder than at building services, as they remain on the critical path to completion for a longer period than any other element.

The most successful jobs will always be those where building site managers have first built teams focused on achieving a successful project. In doing this they must manage the building services elements with the same skills they bring to the foundations, frame, envelope and finishes. In improving project performance in the management of mechanical and electrical (M & E) work, a most difficult area, the builder may bring about, from site level up, a reduction in adversarial attitudes, improved profitability, and happier clients paying less for the product they want. By their efforts, encapsulating the spirit and intent of the Latham Report, site managers can do a great deal for their industry. More importantly for the site managers themselves, they will advance their careers and ensure that they will be of the longest possible duration. Those in training to become site managers, or already running their own sites and seeking continuing professional development (CPD), will derive most benefit from this book. It will also be of benefit to building contracts managers and all those who have a direct and indirect contractual responsibility for building services, such as services engineers working for contractors, services contractors, architects, consulting engineers, project and facilities managers and quantity surveyors.

Acknowledgements

My sincere thanks are due to all of the site, project, contracts and design and build managers, planners, estimators and quantity surveyors who helped distill and define the best practices for managing building services contracts, the greater number of which appear in this book. Most of these personnel, but not all, were colleagues from Wimpey Construction Ltd and its subsidiaries. They are too numerous to mention individually, so a few are chosen to represent the many.

In Wimpey Construction Management and Grove Projects Ltd, Bill Martin and Dick Fry sharpened my focus on management generally. On the building services side, John Kew drew my attention, with alarming clarity, to key issues to be addressed on any project, while Derek Gibson homed me in on areas of risk, and Bob Grice cleared up planning, commissioning and more within the domain of building management systems. Their support and enthusiasm tilled the earth in which the seeds of this book have grown.

Epe Dyer has my gratitude for giving me an appropriate level of computer literacy, thus helping me cure my habit of 'putting rubbish in'.

My thanks to the Directors of Wimpey Construction Ltd for their support and permission to use material which is the source of many of the illustrations.

For the preparation of most of the unassigned illustrations I am indebted to Andy Flint, of Utopia Digital Design, and Bob Grice.

In granting permission to use copyright I am grateful for the support of the Association of Consulting Engineers, N.G. Bailey & Co Ltd, John Berry of Ove Arup and Partners, the Building Services Research and Information Association, the Chartered Institution of Building Services Engineers, the Construction Industry Research and Information Association, Construction Industry Publications Limited, DEGW, Drake & Scull Engineering Ltd, Haden Young Ltd, Hascom Network Ltd, the Heating and Ventilating Joint Safety Committee, HMSO (on behalf of the Health and Safety Executive), Addison Wesley Longman Ltd, the National Association of Lift Makers, and Tolley Publishing Co Ltd.

My thanks must go to Janet Joseph for reading the typescript and converting an engineer's English into tolerable literacy; any residual faults are mine.

To my wife Margaret, who typed the manuscript and provided unflagging support throughout its production, I give my deepest thanks.

Introduction

'M & E' is still the term most commonly used to describe what has become an increasingly sophisticated and widening range of building engineering services elements. They account for a growing proportion of overall contract values. Given their system based characteristics of being constructed by placing on, threading through, and fixing to the building structure and fabric, a bit here and a bit there, they lack the apparent growth and order of the gridline dominated, surface covering, building work. When completed, their ability to sustain a specified internal environment and protect the occupants through guardian systems, must be proven, documented and witnessed, to the satisfaction of the client, designers, authorities, utilities providers and insurers, all within the contract period.

Most books and papers on the management of construction projects – and there are many – have approached the theory and practice of the subject from the company or project corporate viewpoint, the apex of the triangle. These approaches may well have been suited to the static, stage defined, single or limited trade elements of foundations, structure and envelope. However, the documented history of the industry records that many of the management methods it has used were far from adequate. Nowhere are management difficulties more apparent than with building services.

The book's objective is to improve the competence of site managers to manage building services contractors, sub-traders and specialists. Taken in dosage appropriate to project size, services complexity and contract status, the advice is intended to make effective the seemingly 30% of the site manager's time spent dealing with, on average, 20% of the overall project value. The key aspects of building services project plans – programming, including information flow and installation sequence, inspection and testing, commissioning and handover with its training, manuals and record drawing requirements – are dealt with. A framework for competency is provided; the more it is understood through experience and application the more sinewy will it become.

As there is much overlap with the management of any other building work subcontract, emphasis is given to the differences between the construction, commissioning and preparation for the handover of building services systems and the elements of foundations, structure, envelope and finishes. It is assumed the site manager is trained or supported in financial management and less will be said about this in general terms, although specifics will be dealt with.

The book is arranged in two parts. After an overview of building services, Part I looks at how risk is created for the site manager through the parties involved, from the client and professional team to constructors and suppliers, along the chosen contractual route. Risk, in terms of those areas that cause conflict, is examined together with the aspects of services technology that always seem difficult to deliver. The first part puts into context the book's purpose. Part II takes the management of building services contracts from award to handover, and positions it for the building site manager.

There is intended immediacy of application in the structure of the book. In taking the reader step by step through all stages of a building services contract from pre-award to post-handover, help is available at any point.

A common management approach is unfolded that can be applied to building services jobs of all sizes and complexity, irrespective of main and subcontract forms. However, the book covers the key influences of the contractual arrangement upon the way in which building services are managed. From application of the knowledge gained the site manager will be able to work more comfortably with the services contractors through:

- understanding what they have to do;
- helping them organize to do it;
- receiving the evidence (records) that they are doing it to the specified requirements;
- knowledge of its status relative to the programme.

Part One

Building Services

1 An overview of building services

1.1 What are building services?

Building services are engineering systems. They are placed on, threaded through, and fixed to the structure and fabric of a building.

Any building services (BS) system comprises three elements:

- plant
- distribution
- terminals.

Even the most basic building requires six or seven separately identifiable systems to make it work. The building's form and function affect the complexity of building services.

Building engineering services are generally referred to by builders as M & E, mechanical and electrical services. This broad grouping can include public health, fire and security systems. Lifts and escalators are usually referred to under those names and carried out by specialist firms. Other associated but specialist building services systems which may be carried out under separate contracts are:

- sub stations
- high voltage switch gear
- data and telecommunications services
- generators
- uninterrupted power systems (UPS)
- kitchens and cold rooms
- medical gases
- process services.

These services may also be found within M & E contracts as specialist subtraders. We can understand why they are necessary by glancing at Table 1.1.

To support and operate the BS there are the essential utilities:

- gas
- water
- electricity
- drainage
- telecommunications.

Table 1.1 The generic families of building services

Air conditioning
Heating and ventilation
Water services
Lighting
Electrical power
Fire fighting
Smoke control
Fuel systems
Waste systems
Data and telecoms
Security
Special processes
Lifts
Controls
Lightning protection

1.2 Why are building services necessary?

Building services enable buildings to be used for their designated purpose. This they do within a framework of enabling and controlling legislation:

- by creating an internal environment – heating, ventilation, air conditioning, lighting and acoustics;
- by defending the building from the external environment – lightning, rain, wind, noise, heat and cold;
- by providing protection – against fire and for security;
- by enabling communication – through voice, vision and data systems;
- by providing welfare – with toilet and first aid facilities (including those for the disabled), and vending/catering;
- by disposal of waste – through plumbing, recycling and refuse collection systems and services.

Through services systems buildings are made to function safely and healthily.

During building occupation the environmental, power and public health services are dynamic, while those of fire fighting and security are generally passive. The passive systems becoming dynamic only upon activation, which may be by human intervention or automatic sensing. Outside periods of building occupancy the systems are in dormant and passive sensing modes, or in the case of environmental systems maintaining a predetermined lower level of internal climate that can be raised quickly to occupancy standard.

1.3 Matters of design

1.3.1 CLIMATE

Geographically designated as a temperate climatic zone, the weather patterns of the British Isles pose problems for the BS designer. By comparison Scandinavian countries have much lower external ambient temperatures, but the temperature tends to go down and stay down for long periods. This gives an external environmental stability to which internal climates can be matched. But in the British Isles we can be subjected to considerable unpredictable changes in temperature; rain and wind in a 12 hour daytime period. Designing for the 'average' often catches out the ability of internal climate systems to cope with such changes.

1.3.2 ESSENTIAL SERVICES

It is mandatory for all occupied buildings in the UK to have:

- heating
- ventilation
- plumbing
- hot and cold water
- power
- lighting.

and supporting utilities. Certainly the ventilation may be by natural means, via openable windows; it is nevertheless essential in providing the oxygen we breathe.

1.3.3 THE EFFECT OF FUNCTION AND FORM

Services designs are affected by building function, which dictates building form and layout. These latter have an impact on the complexity of the engineering services to be provided and the way they are integrated with the structure, and interface with the buildings fabric and finishes. Some examples of the effect of these aspects on different buildings are as follows:

- A modern hospital of 'nucleus' cruciform design has service streets meeting at intersections where operating theatres, laboratories and toilet facilities increase the density of services provision.
- Manufacturing facilities with process equipment and machinery bring demands for spaces varying from large open production lines with exposed services, to small clean/sterile atmosphere rooms and enclosed services. In either, the mix of process related services may include steam and condense, compressed air, vacuum, cooling water

and drainage lines, clean electrical supplies, high grade lighting, and volatile gases requiring state of the art leak detection.

- The acoustics of theatres and concert halls will make demands on the careful application of the heating, ventilation and air conditioning (HVAC) systems serving auditoria and rehearsal rooms, etc. The rotating machinery of fans and pumps will be of slow speed and isolated from the ducting and pipework systems. Air distribution terminals must be selected to give adequate throw of air without generating noise at the outlet.
- Leisure centres with swimming pools and ice rinks bring specialist complexity to engineering services. Flumes, diving tanks, underwater lighting effects, lighting to avoid spectral glare, water filtration and treatment are all requirements additional to the general services. For an ice rink the integration of the ice pad with the building structure and foundations is an interface requiring particular care.
- For offices the depth of floor plan (shallow or deep), relationship to an atrium, false ceiling and floor depths and the number of service cores will determine the layout of engineering services. Whether it is to be speculative, a prestigious headquarters or a local authority building will determine the standards.

1.3.4 LEGISLATION, CODES AND STANDARDS

The legislated requirements for buildings and their services are very extensive and are treated here in the context of scene setting. Buildings first require planning approval and must be further designed and constructed in compliance with the Building Regulations. Approval to the latter is through local authority building control departments; these may also carry the responsibility for fire approval. Alternatively, fire approval may be delegated to the local brigade. Whatever patterns of controlling organization apply, matters of public and environmental health will be generally embraced by the local authorities. If the building is being procured on behalf of the state e.g. as a prison, government laboratory or defence establishment some 'normal' building regulations may be set aside. But, be assured, they are nearly always replaced by a higher, more onerous level of requirement.

As befits a developed society there is no shortfall, locally or nationally, in the requirements for providing safe and healthy buildings. The overall architecture of relative legislation is framed in the Health and Safety at Work Act etc. 1974. Under this Act, regulations covering premises, plant and machinery, substances, procedures and people have been introduced. One of the most far reaching regulations for the site manager are the recently introduced Construction (Design and Management) Regulations 1994.

For the services designer, compliance with legislation means the acquisition of knowledge so that the system selected to meet the brief are as strong, safe and simple as they can be. Fortunately there is much guidance by way of Approved Codes of Practice (ACoP), Codes of Practice (CoPs) and of course British Standards. The last named are generally recognized as being the minimum standards for components, equipment and system designs. Some also cover system management. Further help is on hand for the designer through membership of professional institutions, with their guides, codes, manuals, standards, technical notes and memoranda. Support can be procured from a wide range of government and industry research organizations:

- the Energy Efficiency Office (EEO)
- the Building Research Establishment (BRE)
- the Fire Research Station (FRS)
- the Building Services Research and Information Association (BSRIA)
- the Construction Industry Research and Information Association (CIRIA).

BSRIA's *Reading Guide 14/95 Building Services Legislation* [1] is a good starting point for any investigation into discovering whether, or what, legislation applies to a subject. For standards, codes, guides and other information available from most of the professional institutions, consultancy associations and learned societies (see Appendix M). Much of the information is available to non-members.

All buildings have to meet minimum standards in the provision of fire detection and prevention systems. Determined by law according to the function of the building they may be enhanced, thereby attracting lower insurance premiums, or backed up because the building must operate at the highest level of availability. Security systems are not required by law but may be provided to enhanced levels for the same reasons as fire systems.

1.3.5 ENVIRONMENTAL IMPACT

Clients, particularly those that trade their products to the general public, are concerned with image and the environmental impact of their buildings. Environmental impact may be looked at on three levels:

- global
- neighbourhood
- internal.

These factors do not always appear in harmony. A new building on a greenfield site may require infrastructure development in the road and public utility demands that it makes. Yet it may be a very good neighbour, being sensitively landscaped and providing jobs. Its location may mean that it can possibly manage without air conditioning. In the case of an office building, the greenfield site may be no better than the town or city centre. The central location and the need to keep out noise, dirt and heat usually make the provision of mechanically refrigerated air conditioning essential; but such a development does not require new roads or services mains. Fortunately the Building Research Establishment Environmental Assessment Method (BREEAM) schemes allow comparison of buildings employing differing BS technologies. Additionally, the EEO offers much good guidance on energy targeting and usage to clients in all sectors of commerce and industry.

It is in the provision of new commercial building stock that we are seeing a greater integration of building services. The orientation of a building, application of shading overhangs and vertical screens, taller, narrower windows with deep reveals, trickle vents, and internally, mass concrete thermal sinks, are all important. They can be used in a variety of combinations to mitigate the need for comfort cooling or full air conditioning.

Building structure and fabric and building services are thus becoming more closely entwined. In such facilities it will no longer be possible to test and prove the building climate services independently of structure and fabric.

1.3.6 SCHEMATICS

The visualization of BS systems first takes place through the designer's production of schematics. Issued to the builder at tender enquiry stage they can be the source of much valuable information. Whether it is generic or specific, a system schematic will show the essential three elements of plant, distribution and terminals. A generic schematic, e.g. Fig. 1.1, will simply show the relationship of the parts for that type of system. Job specific schematics will diagrammatically relate the system selected by the designer to the building's basic geography; Fig. 1.2 shows a small bore heating system for a bungalow and indicates the piping routes from the boiler to the radiators, and names the rooms served.

1.3.7 SYSTEM SELECTION

The generic families of building services (see Table 1.1) do not define the type of lighting, security or air conditioning systems. The designer decides these for the building under consideration and takes into account:

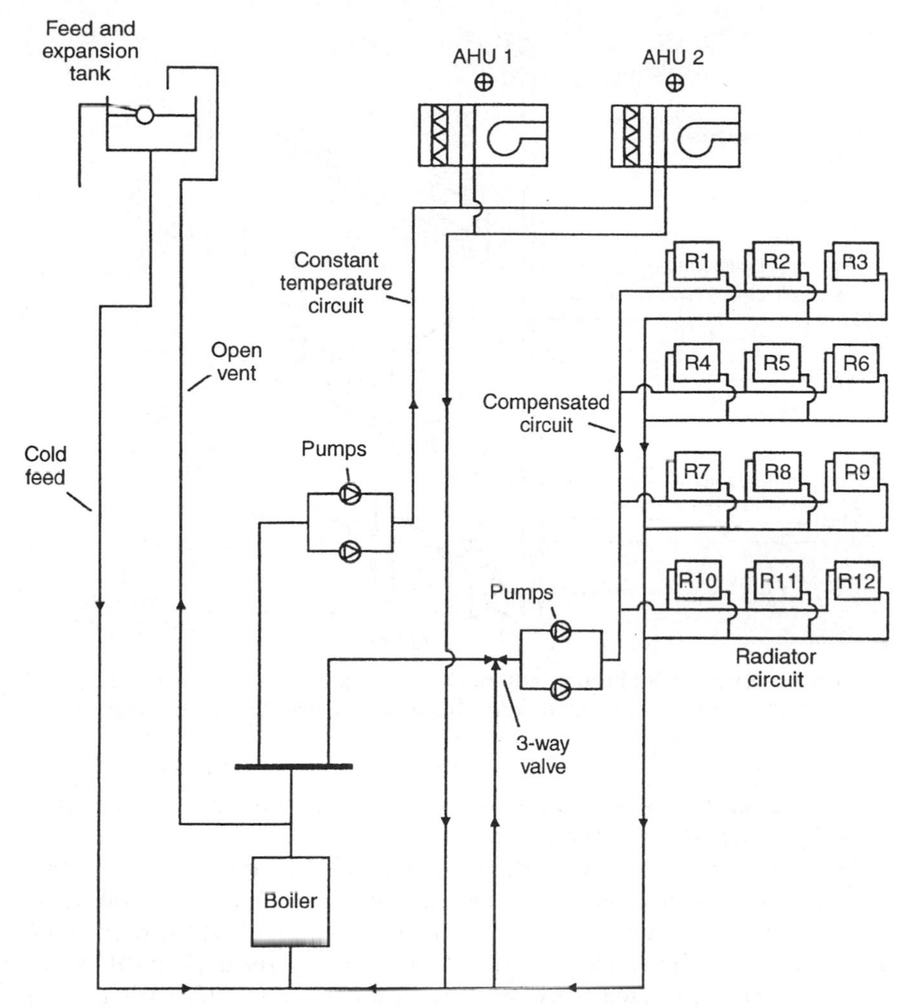

Figure 1.1 Generic schematic of a low pressure hot water heating system. (Source: BSRIA TN 17/92, *Design Information Flow*.)

- capital cost
- running cost
- ease of maintenance
- flexibility (change of layout)
- noise
- appearance of terminals
- space requirements (plant and distribution)
- ease of control
- incursion into usable space
- user acceptability.

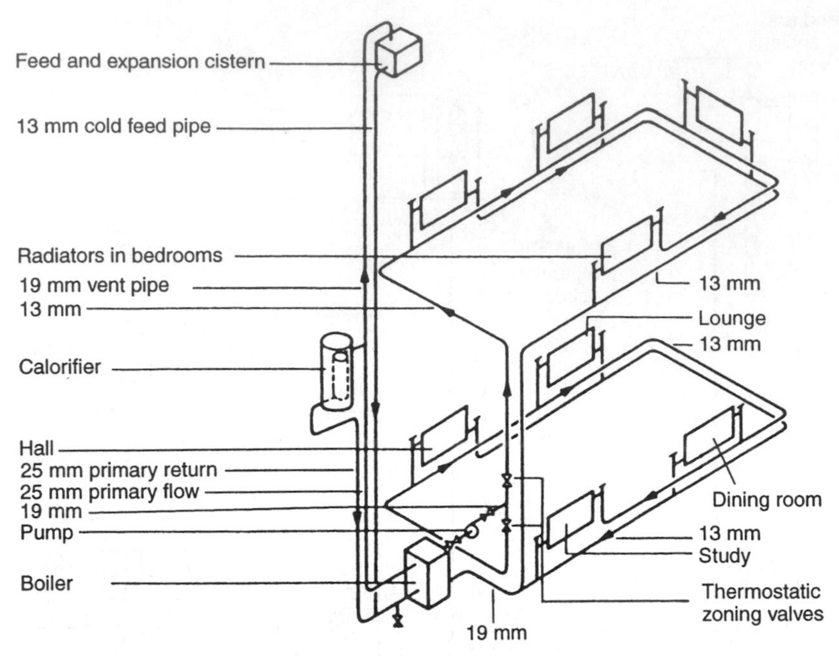

Figure 1.2 A small bore heating system. (Reproduced from F. Hall, *Building Services and Equipment*, Vol. 1, 3rd edn, Longman Scientific and Technical.)

This random listing will vary in hierarchy of importance for each building type and the services within it.

To give an indication of the wide range of building services systems and their subsystems the Co-ordinating Committee for Project Information (CCPI), Common Arrangement Work Section (CAWS) listings R–X are included in Appendix A. The listing takes the form of an alphanumeric reference for each work section. The CAWS are the basis for the Standard Method of Measurement 7 (SMM 7). An indication of the wide choice of systems from which the designer can make selection can be seen under U, ventilation/air conditioning systems, and V, electrical supply/power/lighting systems.

Laid out in a three-level hierarchy from generic to specific type even these are not exhaustive listings. For V 21 general lighting could be provided by tungsten, fluorescent and halogen lamps combined in a similar seemingly bewildering choice of luminaires.

1.3.8 SYSTEM LAYOUT

The design engineer lays out the terminal positions on general arrangement drawings perhaps using computer aided design (CAD). For those terminals on a common system, interconnecting distribution lines will be drawn from the source plant room along distribution routes. The designer will seek, in discussion with the other design team (DT) members, to locate the thermal and electrical power plant in positions which will keep distribution routes as short as possible, while remaining convenient for the connection of utilities.

1.4 Where do they go?

1.4.1 GENERAL

Having given the site manager a basic insight into some aspects of design for M & E we can enhance his (or her) understanding of where they can be expected to be found on the project. A building's function, form and required levels of fire, safety, security, internal climate and reliability, determine the complexity and density of services to be provided. This in turn affects the spaces services occupy and ultimately the size of the building and its overall cost. It is these aspects rather than the sheer size of building served that determine the space given over to BS.

1.4.2 PLANT ROOMS

The size of plant rooms examples the last point. The greater the boiler, chiller, diesel generator required, the more cost efficient they become on a weight and volume occupied basis compared pro rata with units of smaller capacity. In addition the space required around plant items for construction, repair and maintenance seemingly differ very little between the smallest and largest units in a catalogue.

Table 1.2 takes the generic listing of BS and expands it to indicate where the location of major items of plant and equipment are most commonly found.

Figure 1.3 shows (a) the basement plant areas and (b) a section through a prestigious building in which the financial services functions require a high degree of reliability from their services support. The section shows best the take up of space for air conditioning in the general offices and a closer controlled climate for the computer suite. Generator, UPS and Private Automatic Branch Exchange (PABX) make their claim for space. At roof level there is equipment to reject heat from the air conditioning systems, water storage and air handling unit plant rooms, lift motor room, aerial and satellite arrays. The plan depicts the loss of floor area due to vertical transport systems and interconnection between chillers and condensers; with these we are starting to move away into the general distribution routes.

Table 1.2 Location of plant and equipment

Services	*Major plant items*	*Possible locations Low/inter/high level/ External/other*
Air conditioning	Refrigeration	Low/high/external
	Cooling towers	High/external
	Air cooled condensers	High/external
	Pumps	Low/high/external
Heating and ventilation	Fuel store	Low/external
	Boilers	Low/high/external
	Flues	As boiler location
	Pumps	As boiler location
	Air handling units	Low/inter/high
	Hot water storage	Low/inter/high
	Pumps and panels	Related/to storage location
Lighting and electrical power	Substation (transformers)	Low/external
	High voltage (HV)/low voltage (LV) switchrooms	Low
	Generators	Low/external
	UPS	Low
Fire fighting	Hosereel water storage and pumps	Low/external
	Sprinkler control room/ Water storage	Low/external
	Wet riser storage and pumps	Low/external
	CO_2 and foam	Near space served
Smoke control	Fans	In or near space served
Fuel systems	Solid/oil/gas	Low/external
Waste systems	Sewage holding tanks	Low
	Sewage and sump pumps	low
Data and telecoms	Exchanges	low
	Frame rooms	low
	Equipment rooms and closets	Low and all intermediate
	Satellites and aerials	High (roof)
Security	Monitoring station (control)	Low
Special processes	Plant and Equipment for compressors, vacuum pumps and generators	Low/external/near served area
Lifts	Electric traction M/C rooms	Above/below shaft served
	Hydraulic ram	Borehole below car, pump room adjacent
	Hydraulic side acting pumps	Low near shaft

Table 1.2 *Continued*

Services	*Major plant items*	*Possible locations Low/inter/high level/ External/other*
Control room	Energy, fire and security building management monitoring station	Low

1.4.3 UTILITIES

Before considering internal distribution let us go back to Table 1.2 and unravel the space and location requirements for the utilities. Following the privatization of water, gas and electricity, supply companies are now happier to contract for providing a service into a building. Whereas previously, even for sites with no more than a few metres from boundary to building, the utility service would be terminated and metered at the boundary, now privatized companies are not only happy to come into the building but may seek to contract for carrying out internal services. Nevertheless, gas, water and electricity metering and intake rooms are best provided near to ground level, preferably with controlled access from the external face of the building. This is particularly so with respect to electrical supplies where, in the case of sharing

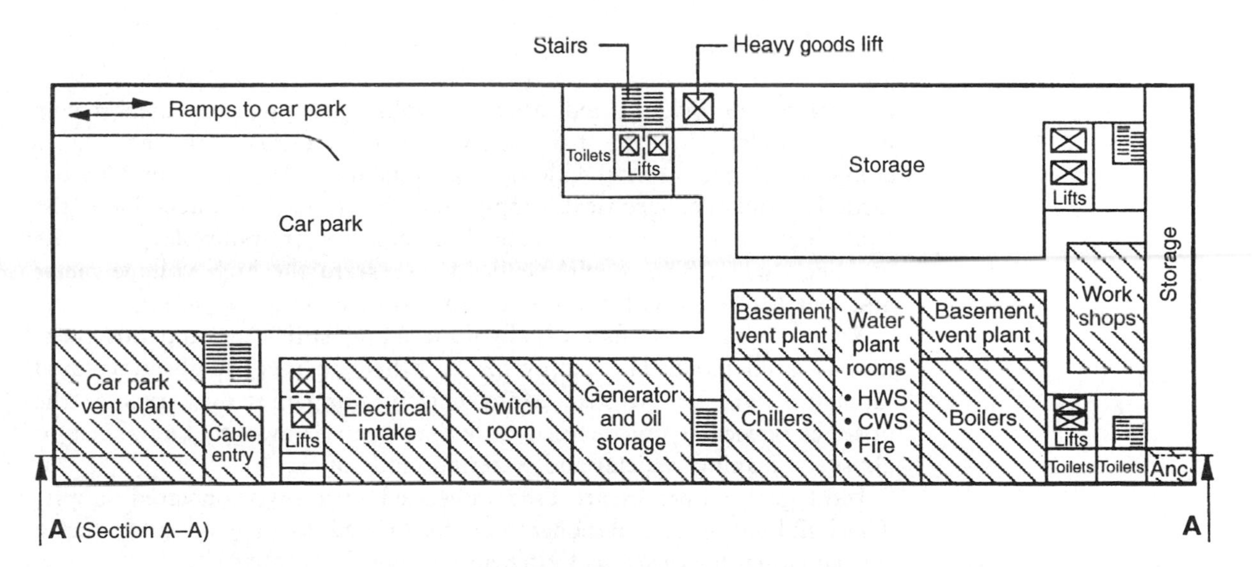

Figure 1.3(a) Prestigious commercial offices, basement plant areas.

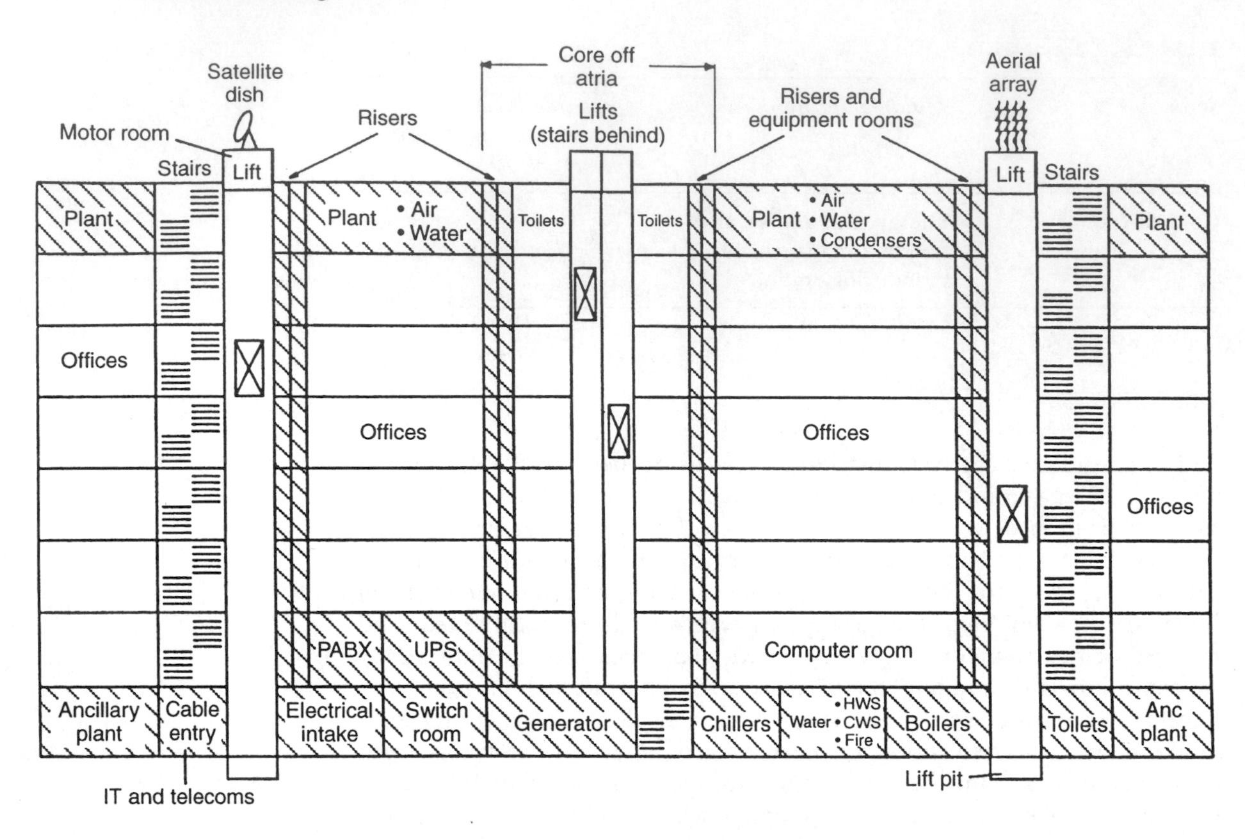

Figure 1.3(b) Prestigious commercial offices, section A–A and plant and risers.

an area substation or a dedicated low voltage supply, permanent access to the intake room must be granted with wayleaves to the supply company. If the size of building warrants it, and a more preferential tariff is available, electrical supply may be at high voltage. Here the substation and its maintenance becomes the responsibility of the building owner. The requirements for access to the high voltage meter and switch remain with the supply company, as with low voltage.

Water supplies for fire, e.g. hydrant main, sprinklers and hosereels, may be unmetered. The supply for all other services usually designated 'domestic' will be metered. Many supply companies now require the hosereel service to be metered as it has been known to be subject to abuse in cleaning vehicles and watering landscapes.

Most gas companies are only interested in a single metered supply. The building services designer may be briefed to provide submetering for individual tenancies and kitchens.

The deregulation of communications has also increased demand on building space for engineering services. Previously one only had to consider British Telecom; now we have Mercury and others.

It is usually possible to design the rainwater and foul drainage installations to gravitate from the building. Where this is not possible, soil and surface water drainage can be collected in chambers and discharged through pumped mains into the local authority systems. Normally, sewage pumps are duplicate sets arranged for cascade back up operation.

1.4.4 DISTRIBUTION ROUTES

We return to consider the distribution of services between and through buildings. There is a hierarchy of level to be considered. Primary distribution takes place from plant areas both horizontally and vertically. Common route types are, horizontal – crawlways, ducts and trenches and the corridor ceiling void – and vertical – risers (multistorey).

As Fig. 1.4 shows, within a heirarchy of primary and secondary distribution routes there can be considerable geometrical variation. The example could apply to ventilation distribution from an air handling unit, heating pipework from a roof level boiler plant, or water distribution to laboratory benches.

Figure 1.5 is an example of vertical primary distribution within a city centre air conditioned office building. Homing in on the toilet block core in Fig. 1.6, it is seen to be encased by the vertical distribution of nearly every conceivable service system for a building of that type.

Secondary distribution in most types of buildings is arranged horizontally. In some buildings such as hotels and vertically stacked toilet blocks, the connections between risers and terminals are short. Here the concept of secondary and tertiary distribution becomes blurred, but in many buildings, particularly offices, secondary distribution takes place in either floor and ceiling or in both. Figures 1.7 and 1.8 show secondary distribution taking place mainly in the ceiling and floor respectively. In these examples of secondary distribution through floor and ceiling voids it is seen that the terminals are located on the surface of the false floor and ceiling, i.e. the tertiary distribution takes place in the same space. This occurs most commonly with lighting, ventilation or air conditioning systems and to a lesser degree with sprinklers and fire detection heads.

Tertiary distribution is defined as a situation where services are taken from either the secondary distribution or boundary of the serviced space to some point within it. The method of distribution such as dado or skirting trunking, service poles and rails usually takes up some room space. In highly serviced buildings or for aesthetic reasons this final services distribution can be integrated by designers into the furniture

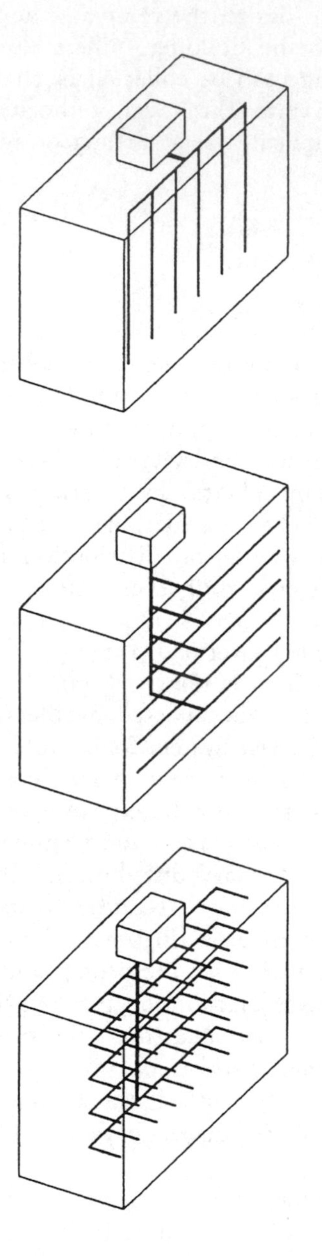

Figure 1.4 Primary and secondary distribution routes – geometrical variation.

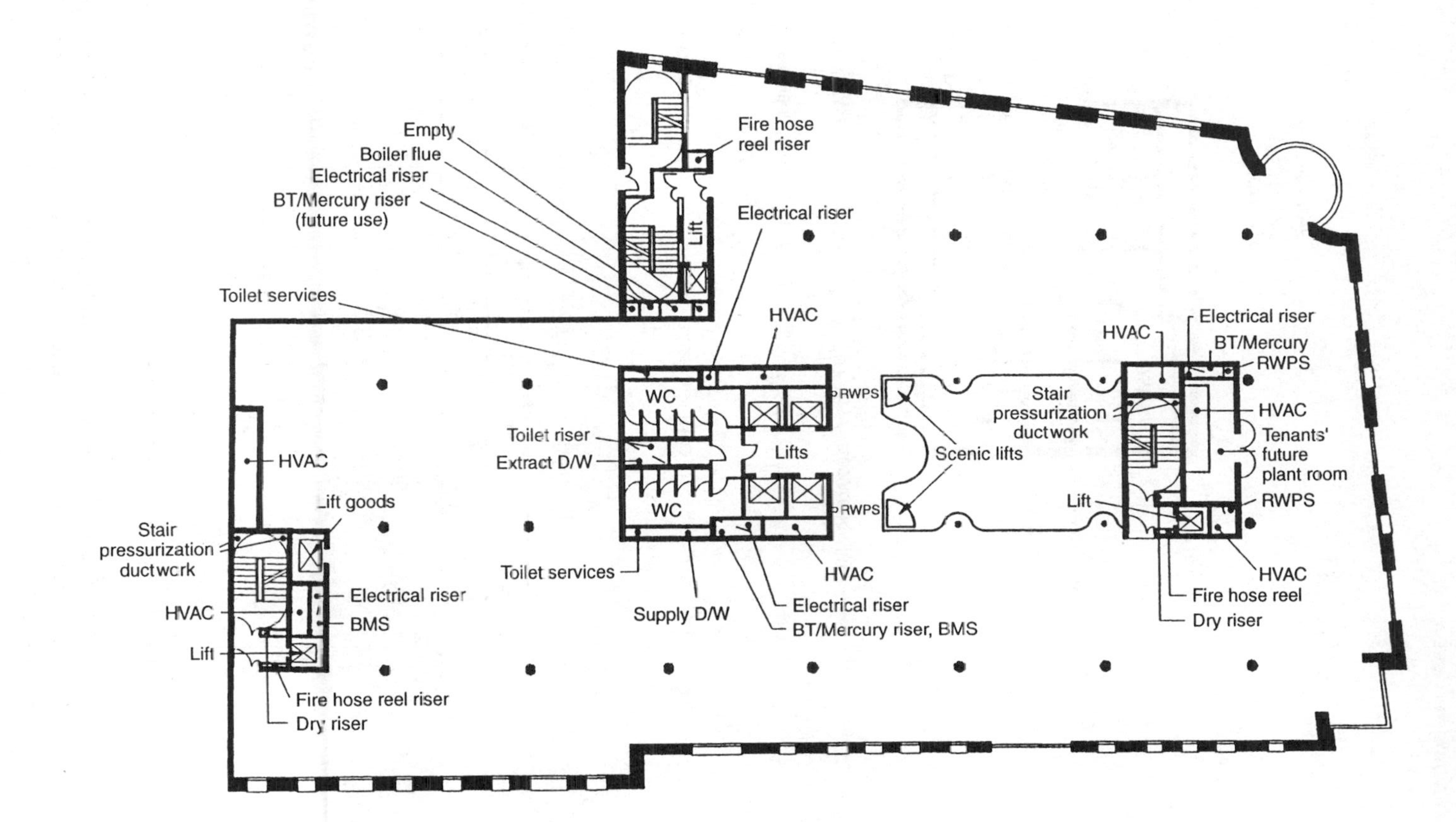

Figure 1.5 Vertical primary distribution spaces within a city centre air conditioned office building.

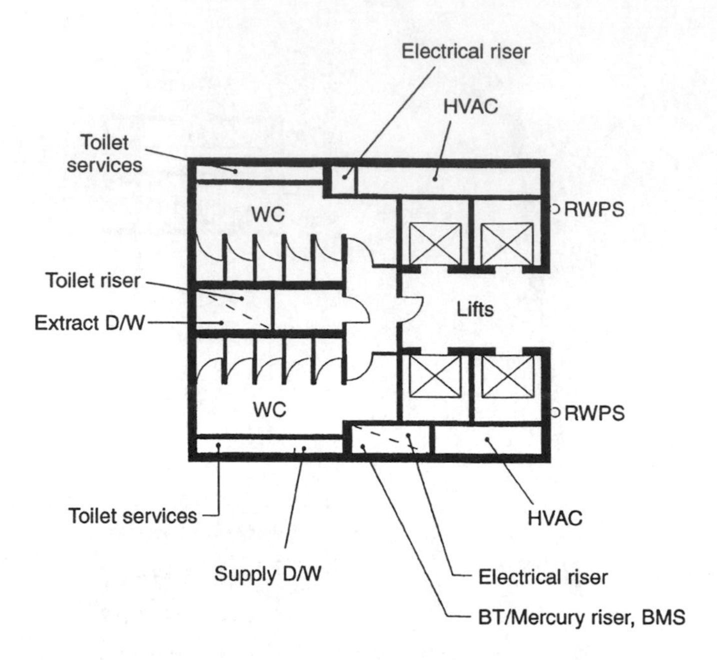

Figure 1.6 Toilet block core – vertical services distribution spaces.

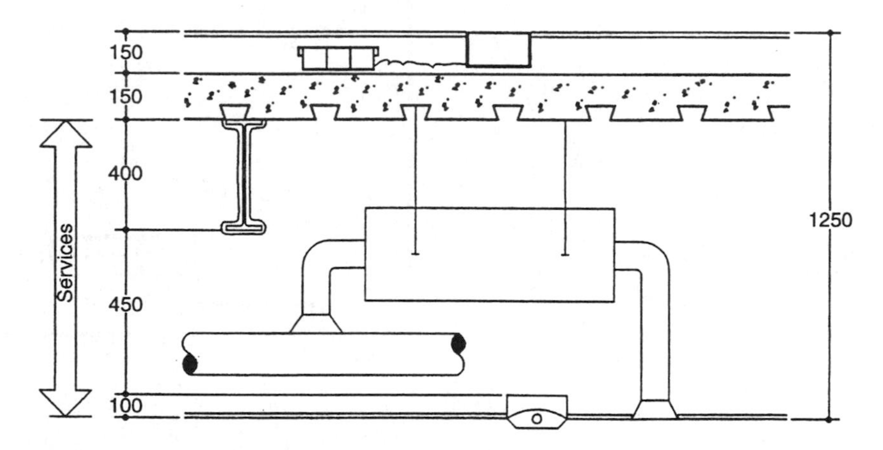

Figure 1.7 Secondary distribution in ceiling: steel frame metal deck, small raised floor, full access false ceiling, recessed light fittings. (Source: J. Berry, Ove Arup and Partners.)

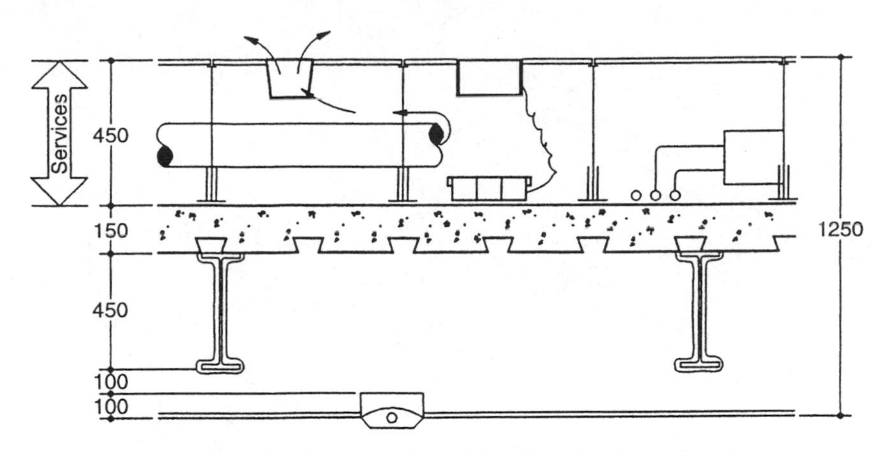

Figure 1.8 Secondary distribution in floor: false floor depth increased, services removed from ceiling, zone-transferred to floor zone. (Source: J. Berry, Ove Arup and Partners.)

and partitioning. Figure 1.9 shows servicing at the perimeter and screen wall, and Fig. 1.10 workstation servicing in lay-in ducts and hollow section risers.

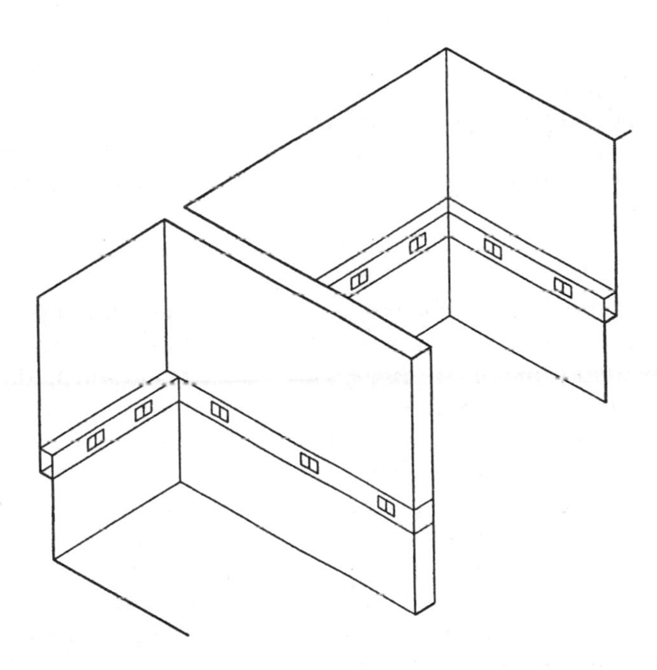

Figure 1.9 Tertiary distribution – servicing at perimeter and screen wall. (Source: DEGW.)

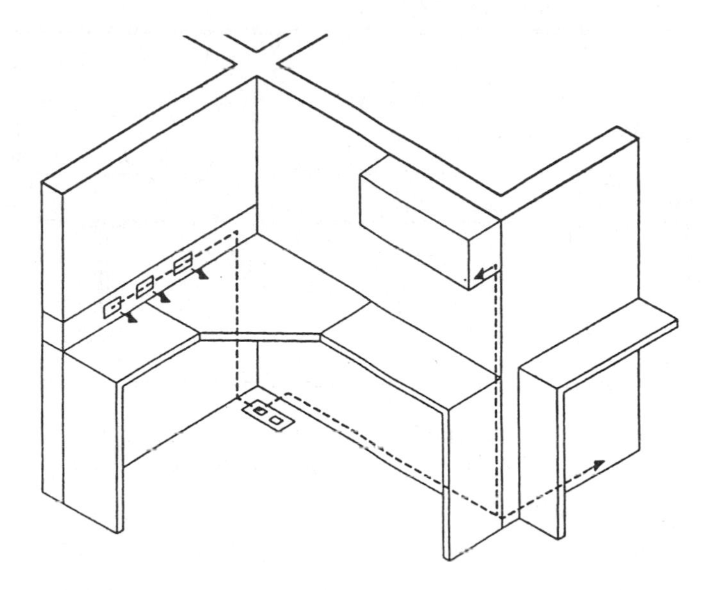

Figure 1.10 Tertiary distribution – workstation servicing in lay-in ducts and hollow section risers. (Source: DEGW.)

Terminals

The terminals of M & E systems are therefore to be found in walls, floors and ceilings, and the fit out furniture of e.g. offices, laboratories and hotels. They may be surface mounted, semi- or fully recessed.

Industrial buildings

In the large open space of factories, warehouses and DIY superstores the distribution of building services is generally exposed. The main distribution takes place parallel to eaves and valleys from which there are a variety of routes to the terminals. Where required by legislation or insurance, sprinklers may run parallel to the pitch of the roof. At intervals electrical power will follow this slope to smoke vents and openable roof lights. Both of these may alternatively be activated by compressed air. Depending on the form of structural frame, truss or portal, secondary support grids may be required from which will be suspended ducting, piping and power and lighting distribution to fittings suspended to create a notional horizontal plane at eaves and valley level.

Piped services to process machinery featuring steam and condense, compressed air, vacuum, gas, treated and chilled water, can be run in various ways. Commonly, they will drop adjacent to structural columns and run in floor ducts to the production lines. For steam or water 'used once', or even where process heating or cooling water is closed circuit, sumps, possibly with pumps, will be provided with an underslab gravity drainage system. The piped services may also be provided via overhead tracks with drops to the machines. These may be encased in service poles.

1.4.5 FIRE SERVICES TO BUILDING CORES

As with the plumbing and water services obviously associated with toilet areas, there are other services to be found in building core areas. For economy of building layout enabling common use of circulation space, toilets in multistorey buildings are found arranged with lift lobbies, disabled refuge areas and staircases. Contained in an appropriately fire rated compartment, it is here that we will find break glass unit fire alarms and hosereels, wet and dry fire risers. The staircases and lift lobbies are increasingly being required to be pressurized against smoke ingress.

Fire and acoustic transmission paths

Wherever building services pass through an element of structure or building fabric there is a risk of creating a fire and noise transmission path. Should any of those penetrations occur in a fire rated element it must be sealed to maintain the integrity of the fire compartmentation.

Both the requirements for fire compartmentation and the methods for compliant 'fire stopping', where services pass from one fire compartment to another, are given in the Building Regulations. Many of the methods available to satisfy these regulations will also be effective in preventing or minimizing noise transmission.

1.5 Who designs them?

1.5.1 GENERAL

From the trunk of generic families of BS listed in Table 1.1 there are many branches. There are literally hundreds of different types of BS systems. There are forty to fifty derivations under air conditioning. The constant churn of technological advancement in lighting, voice, vision and data services seems exponential in its progression curve. Everywhere the application of microprocessor technology aids these advancements. What does all this mean for the BS designer? The need to acquire, disseminate and apply this knowledge means there are few BS consultancies or designer contractors with capabilities to design every building service in house.

1.5.2 CONSULTANTS

Consultants do not design lifts, escalators, transformers or generators, equipment or terminals. However, from their knowledge of the function of these components they are able to select and integrate them into services systems for incorporation into buildings. The situation is similar with respect to sprinkler installations: few consultants have the capability or, more importantly, are certified to carry out the complex hydraulic calculations in compliance with the 29th edition Loss Prevention Council (LPC) *Rules for Automatic Sprinkler Installations* (now concomitant with, BS 5306) [2].

For all services, the consultant engineer must be capable of defining the performance criteria and giving weight, and spatial requirements to the structural and architectural members of the design team. To do this it may be necessary for the consultant to obtain information and take advice from the plant and equipment manufacturers.

Prior to 1995 most engineering services design appointments were made under, or derived from the Association of Consulting Engineers three Schedules of Duties. Known as Appendices 1, 2 and 3, they rose in levels of responsibility from 'Performance', through 'Abridged' to 'Full duties' respectively. Building services technology, limits of harmonization with the RIBA Plan of Work, the use of a wider choice of contractual routes and politics outpaced the usefulness of these schedules. The hard market conditions of the early 1990s and political calls for fee competition increasingly exposed their limitations. They have now been replaced by the Association of Consulting Engineers (ACE) Conditions of Engagement 1995, Agreements A(2), B(2) and C(2) [4]. These allow the knowledgeable client to list those duties he wishes the designer to carry out. In other cases, and these may include fee bids, the designer will list duties and a fee for their discharge. In both cases you get what you pay for and there is an onus on the design procurer knowing what is needed; see section 2.2.3, 'Terms of engagement'.

There is an aid to the assignment of duties. The BSRIA has published Technical Note TN8/94, *The Allocation of Design Responsibilities for Building Engineering Services – a code of conduct to avoid conflict* [5]. This provides pro formas for more closely defining those areas of design responsibility that, through differing interpretation, regularly lead to dispute between the designer and the installer (see section 2.2.4, 'Division of responsibilities').

1.5.3 THE DESIGNER – CONTRACTOR

The reader will have noticed the use of both terms 'consultant' and 'designer'. Not all design is carried out by consultants or specialists under their control. A considerable number of BS contractors undertake design in a variety of ways:

- as designer installers working for design-and-build main contractors;
- appointed by developers and other end user clients to work with the separately appointed professional design team;
- appointed by the lead professional designer, usually the architect;
- appointed by project managers to work with the professional team.

There are a number of variations, but all have some direct line design warranty to the end client.

1.5.4 NOVATED DESIGN

A further design procurement variant quite frequently used is that of novation, often found in the main contract design-and-build route. Here a consultant may produce a performance or abridged duties design which is passed via the design and build (D & B) contractor to the services designer–installer. These may be novated with a requirement to accept full responsibility for the design and its further development. Sometimes the employer's requirements ask for the D & B main contractor to take the consultant under contract. In turn this leads to the main contractor seeking to mitigate his design responsibility by bringing the consultant and designer–installer together with a form of 'back to back' design warranty.

1.5.5 THE INSTALLER

Historically, the greatest percentage of BS design work has been undertaken by consultants working to the ACE 'Abridged duties'. Reduced to its simplest terms this comprised a specification and set of drawings. These last named took the form of general arrangement drawings at 1:100 scale. Some plant room layouts were to larger scale – 1:50; for toilet block layouts, 1:20 and perhaps a few sections depicting the preferred arrangement of services in risers, crawlways, ducts, trenches, ceiling and floor voids. Some consultants enlightened as to where 'pinch points' would occur at congested intersections along the distribution routes would produce better details to show the viability of installation. At best these schemes on 'Abridged duties' were only numerical solutions. Through the development of working drawings and at the workface of construction the installer proved whether they would work.

That now defunct major client, the Property Services Agency (PSA), under the spotlight of scrutiny in spending public money, could not be seen even via its agents, the design consultants, to bestow favour upon one manufacturer or supplier in preference to another. The PSA's consultants were not allowed to obtain competitive quotations, or to preselect plant, equipment or terminals. This led to a certain imprecision in designs for the PSA. There are still many clients appointing consultants who will not let them preselect on 'Abridged duties'. The installer – and remember he is not seeking to trade on his design knowledge – bids the material content for the contract by preparing estimates for plant, equipment and terminals, scheduled by capacity/performance only. Having won the contract the installer sets about producing working drawings. Working with better detail than the consultant on structure, cladding, brickwork, floors and ceilings, etc., the installer will naturally seek economical routing and fixings. Notwithstanding that reasonable objective he will find the need for more bends and sets in his distribution systems than the consultant envisaged, when working with less detail. The consultant, being aware of this, would have called upon the installer to calculate the final air and water circuit resistances to flow, for sizing fans and pumps. The picture is emerging of the extent of 'design' knowledge that the installer must have. More than in any other construction trade the installer's working drawings, offsite construction and work face practices can affect the consultant's 'Abridged duties' design intentions, via:

- fixings;
- anchor points;
- take up of expansion;
- gradient of pipework (venting and draining);
- change sections on ductwork;
- the routeing of electrical conduit;
- the interpretation of earthing and bonding;
- quality of system preparation (cleaning ducts, flushing and chemical cleaning of pipework);
- offsite validation of software.

1.5.6 THE DESIGN ENGINEER

The designer has been referred to as one person. Obviously, this not the case. Whether or not it is theoretically possible for one individual to be capable of designing every building service system, that person has yet to exist. From an early point in an engineer's academic and industrial training, personal preferences come to the fore, and trainees set out to become a sprinkler, lift, electrical, controls, plumbing or HVAC engineer. Some engineers encompass a wider range of building services design than others and go on to be called mechanical or electrical engineers. Most start from some specialist base. All are BS engineers and according to their function and status you may find a mechanical, electrical or public health engineer with overall responsibility for coordinating the design of a project's services.

1.6 What do building services cost?

1.6.1 AS A PERCENTAGE OF OVERALL PROJECT VALUE

There are lies, damn lies and statistics and then averages. What follows is in the last category. The figures in Table 1.3 are averages. Those who wish to denigrate them by quoting their latest, or most recent project experience with services values outside these ranges will certainly be able to do so. It is considered that the percentages may be of use to the site manager who, by calculating both upper and lower percentages, will arrive at a capital cost range. If the site manager's project falls significantly outside that range he would be advised to seek some understanding as to why. The percentage figures are to be applied in calculating the value of services inside the building and must be related to the overall cost of buildings. A great number of quantity surveying practices are very skilful at cost planning external works with their infinite variables for city, out of town, hard and soft landscaping permutations. But even the most skilful of QSs can fall foul of the unpatterned costs for utilities connections, mains network reinforcement costs, and that great catch all, contribution charges. For all these reasons it is inappropriate to give any worthwhile assessment of what external costs may amount to. In defence of the bet hedging width of

Table 1.3 Building services costs as percentage of overall job value (internal services only)

		%
Offices	Natural ventilation	20–30
	Air conditioned	25–35
Warehouses		15–20
Shopping centres		15 +/–5
Factories	7.5 shell:	20–30 working
Leisure centres		20–40
Hospitals		30–50
Motels		20+/–5
3/4 star		30–45
Computer centres		50–60+
Social housing		15–25
Student accommodation		15–20

the percentage ranges quoted in Table 1.3 note the following commentary.

- *Offices*. With the increasing use of building design to minimize the worst effects of our climate, the cost of services in these buildings may fall to around or below 20% of the project value. But beware, we are moving into an area where BS and building works can become blurred. Similar thinking can be applied to air conditioning, where building structure such as reinforced concrete high mass thermal sinks, dense building envelope fabric and/or some screening, reduces the size and cost of air conditioning systems. Some of the newer variable refrigerant volume (VRV) systems, usually of Japanese origin, may be applied as comfort cooling systems. Then the percentage cost for an air conditioned building may fall even further, from 25% to 20%.
- *Warehouses*. If they are simple with few workers and the stored product does not require in-rack sprinkler systems then 15% is about right. This may also apply to basic services in DIY stores. For warehouses requiring a more sophisticated environment, high bay lighting of good colour rendition, fire protection and detection, with security systems wiring and battery charging for automatic guided vehicles, 20% is nearer the mark as you approach the bottom level figure of a working factory.
- *Shopping centres*. Here 15% will cover the provision of landlord's services to the public areas, i.e. malls, toilets, stair and lift lobbies, and car parks. It will also cover services along trucking routes and into each unit, for extension and fit out by the tenant. The 15% plus or minus 5% can also be applied to multistorey car parks.
- *Factories*. Services to factory shells and small industrial units are around 7.5%. When you get into working factories the range is quite wide with the upper figure probably being capable of providing some services connections to the machinery.
- *Leisure centres*. At the lower end of the range we have the 'dry' leisure centre, i.e. sports hall with a number of smaller multifunction rooms, gymnasium and snooker rooms. At around the 40% value mark we are extending from the 'dry' into leisure pools with their flumes, diving tanks, water filtration and treatment requirements and on into the truly multifunctional leisure centre which also incorporates an ice rink.
- *Hospitals*. At around 30% we are covering the provision of major hospital extensions, while new district general facilities can absorb towards 50% of the project's cost.
- *Hotels*. Motels can be serviced for around 20% of the overall value. Three- to four-star hotels if out of town will cost about 30%, possibly up to 40% with conference and leisure facilities. For the

same level of accommodation in town and city centres with conference and leisure facilities plus a greater variety of speciality restaurants, etc., costs can rise to 45% and above.

- *Computer centres*. For those dedicated facilities located out of town on secure greenfield sites operating possibly as 'back up' to the 'back up' centre of financial institutions, services costs can easily account for 60% plus. This high figure is caused by the need for 100% reliability involving standby generation, UPS, high security, high fire safety, duplicate pumps, fans and standby everything.
- *Social housing*. The wide range is affected by the mix of accommodation units and whether they are provided with low capital electric heating or the higher cost wet systems (radiators/fan convectors).
- *Student accommodation*. This follows the pattern of social housing for similar reasons.

To all of these cost indicators must be added lifts and escalators, where the cost varies according to whether the building is low or high rise.

1.6.2 COST RATIOS

In the same way that averages exist for overall building services costs, so also there is a crude pattern of cost relationship between the building services. This applies for a wide variety of buildings, the main affecting variable being that of air conditioning. These ratios which are most accurate for office buildings are shown in Table 1.4.

Table 1.4 Approximate cost ratios for offices (internal services, excluding underslab drainage)

Natural ventilation	Air conditioned
Electrical = 50% of mechanical	Electrical = 40% of mechanical
Public health = 50% of electrical	Public health = 40% of electrical
Example (5000 m^2 offices)	
@£150/m^2 = £750 000	@£300/m^2 = £1.5 million
Mechanical = £430 000	Mechanical = £970 000
Electrical = £215 000	Electrical = £380 000
Public health = £105 000	Public health = £150 000

Table 1.5 Building services costs material/labour ratios

Mechanical	2.5 to 5:1
Electrical	1 to 2.5:1
Public health	1 to 2:1
Sprinklers	1:1
Lifts/escalators	10:1

1.6.3 MATERIAL AND LABOUR RATIOS

The relationship that holds up well is the ratio of material, and here we are including sublet trades, i.e. ducting, insulation and controls, to the cost of labour. These ratios are shown in Table 1.5.

- *Mechanical services.* Where these are predominantly piped services then the ratio of material and labour can be found around 2.5 to 1. Where there is a greater use of offsite manufacture such as in air conditioning with its appetite for sheet metal ductwork and large pieces of plant like chillers, air handling units, etc., then the material to labour ratio widens steeply.
- *Electrical services.* The ratios are closer here, increasing at the upper end due to large pieces of equipment or major sublet items, e.g. diesel generators and HV substations.
- *Public health.* An even closer range than for electrical services due to the high predominance of pipework services. Where the cost of sanitary fittings is included the materials/labour ratio edges upwards.
- *Sprinklers.* The extensive use of labour intensive pipework holds the ratio at around 1 to 1. The use of prefabricated piping can increase this ratio.
- *Lifts and escalators.* The assembly on site of major components manufactured offsite gives these items the highest ratio of all.

1.6.4 LABOUR RATES

For 10- or 11-man gangs, obviously quite large projects, all inclusive but unprofited labour rates can be found in the order of:

- mechanical services: £8–9 per hour;
- electrician: £9–10 per hour;
- ductworkers: £8–9 per hour.

In a period of relatively low inflation (1996) it is impossible to predict for how long these figures will remain valid. In addition to these there are of course many specialist trades for which other rates will prevail, e.g. high voltage electrical workers, controls technicians, commissioning engineers and data and telecom installers.

1.6.5 M & E SERVICES PRICE BOOKS

A great deal of this information on costs has been gathered over time and through analysis and application of Spon's *Mechanical and Electrical Services Price Book* [6]. Published annually it is of the greatest value in acquiring familiarity through regular use. This and similar works are commended to the site manager so that he can check the value of services, their cost and material-to-labour ratios on every job. The site manager who does this will acquire a useful 'feel' for the correct level of project services costs.

References

[1] BSRIA (1995) *Reading Guide 14/95, Building Services Legislation,* Bracknell.

[2] The Loss Prevention Council (LPC) and British Standard BS5306: Part 2: 1990 *Fire Extinguishing Installations and Equipment on Premises* (from the Insurer's LPC *Rules for Automatic Sprinkler Installations*).

[3] The Association of Consulting Engineers (1981) *Conditions of Engagement,* Agreement, 4a for Engineering Services in Relation to Sub-contract Works, amended 1990–1993, London.

[4] The Association of Consulting Engineers (1995) *Conditions of Engagement 1995*, Agreements A[2], B[2], and C[2], London.

[5] BSRIA (1994) *The Allocation of Design Responsibilities for Building Services – a code of conduct to avoid conflict*, Technical Note N 8/94, Bracknell.

[6] E & FN Spon (annual) *Spon's Mechanical and Electrical Price Book*, London.

Risk and its mitigation 2

2.1 Overview

2.1.1 GENERAL

More than anyone else the site manager will appreciate just how risky the construction of a building can be. Every day the site manager picks up and shoulders the burden of risk, much of it created by the decisions and actions of others. Many of these risks are passed on by others in mitigation of risk to themselves or those they represent. In this chapter we will look at what passes down the funnel of risk that, despite its disposition on site, still requires the site manager to deliver the project to the specified quality, at tendered cost, and of course, on time. Specifically we are looking at how risk affects the ability of the site manager to manage the BS contracts.

In this examination of risk and its mitigation there is much that is, or should be, understood about the project by the time we are approaching the granting of the BS contracts. Analysis of the key causes of risk show that they apply to the project overall and are therefore pertinent to every building element or package of work placed on order. These common factors which will be looked at are those born of the client and design team appointment relationships and the selection of the contractual route. The particular risks in services lead from the designer's duties under the terms of engagement to how appropriate and clearly specified are the building services systems. It is from technical matters that most unintentional and therefore unknown risk springs.

The subject of risk and its mitigation is so complex that it is dealt with here on a 'need to be aware' basis.

2.1.2 KEY CAUSES OF RISK

Lack of success in the way all parties involved in a project perceive and manage risk will manifest itself on site in some level of conflict. Table 2.1 is a simple two source categorization for the key origin of conflict. The first source column represents the client and design team, the second column the contractor/subcontractor. Some aspects have their origins in both columns. There are other aspects that would make this table even longer, but being of less importance would serve no purpose

Table 2.1 Causes of conflict

	Aspect	Source	
		Client/DT	Contractor/ Subcontractor
1.	Design fee competition leading to the consultants not working in the client's best interest and providing a reduced service that serves no one well	*	
2.	Inappropriate contracts rarely used in standard form. Perceived as having too many loopholes, tweaked to death by quantity surveyors, scrutinized by the building contractor and given a further tightening in subcontract form	*	*
3.	Services consultants specifications, preliminaries and preambles – regrettably in the past too often taken as an opportunity to rewrite the contract/subcontract by those not party to either	*	*
4.	Coordination of services, to structure and to fabric so ill defined that you can choose your own interpretation as to the level of responsibility for designer and services contractor. The interpretation of expectation by each of the other overlapping by an extent indicative of the shortfall of the service from each	*	*
5.	Tendering is too costly by too many on the bid list. Periods are too short and the prices too low	*	*
6.	Claim consciousness precipitated by need to clawback low tender prices means underpricing of staffing, operating and maintenance manuals, record drawings and commissioning. Contracts are studied with a view to making claims	*	*
7.	Offers of alternatives by contractors putting forward cheaper manufacturers who do not have test or design back-up resources. These prices are low enough for the client's quantity surveyor to force the DT into costly evaluation of the products. These evaluations would not have been allowed for in the design fees	*	*
8.	Programming. Resistance by building contractors to give other than start and finish dates. They are afraid of providing details for the sequence of building works required prior to services, as the missing of those dates could form the basis of a claim. Designers are hampered by the lack of dates when information is really required		*
9.	Installation problems. Services don't fit, fixings impossible, design unsuitable for phased handover	*	
10.	Commissioning. The compression of time due to construction delays in earlier elements. Lack of knowledge of the commissioning process. Systems uncommissionable		*
11.	Systems do not work. Installed and commissioned to specification, they prove incapable of meeting design criteria	*	
12.	Poor payment. Turning the search for reasons of non-payment into an art form	*	*
13.	Communications. Arising from any combination of the above, problems can degrade a contract into the writing of excessive correspondence much of which is carelessly written and even more carelessly read in anger	*	*

in the context of covering the origins of conflict. The categorization of conflict into two sources may be criticized as being an oversimplification. Nevertheless it is the framework within which the detail of risk can be studied further.

A count of the key causes almost shows the same number originating from client and design team decisions, as from contractor/subcontractor actions and responses. It is of no comfort to the site manager that both groups can turn in self-justification of the pain they are causing to the prevailing pressures of the country's economic climate. In boom times contracts are softer, fee and tendering margins comfortable and generally a more harmonious atmosphere pervades the project. Conflict still exists but there is more willingness to resolve it without resort to litigation. In recession survival is everything, contracts are hard and tendering levels subeconomic, breeding risk which results in company failure and a high level of litigation. All of this is 'the norm' – for the moment, as is any economic climate between those extremes.

2.1.3 TECHNICAL

A project may fail technically in an infinite number of ways. Only the most common of these can be looked at here by way of offering guidance to the site manager on the need to understand the building services technology within his project and points of potential failure.

2.1.4 THE CONSTRUCTION (DESIGN AND MANAGEMENT) REGULATIONS 1994

Without a doubt the obligation to comply with these regulations adds to the workload of the site manager and his team. The regulations are nevertheless brought in to identify, obviate, control and thereby mitigate risks to the project in matters pertaining to health and safety under design, construction and management. On completion of the project the building owner or tenant is given a health and safety file. Together with the operating and maintenance manuals and 'as built' drawings the owners/occupiers have the wherewithal to properly interface their responsibilities as employers with those of running the building as a safe place of work.

2.2 The client and design team

2.2.1 GENERAL

Every client wants to know as soon as possible what the project out turn cost will be. It is doubtful whether there has ever been any building project in which cost has not been important, particularly since the Industrial Revolution. To the client project cost means viability, whether it is for constructing a place of worship, a shopping centre, a hospital or the ubiquitous office block. Since the 1980s through the impact of privatization and the not unrelated boom in facilities management, there has been a sea change in perceptions of building cost. Clients with structured property and services management departments have become more knowledgeable about influences on capital and operating costs. For example, in purveying services or goods to the public a more than superficial image of greenness will count, as will rising standards and legislation. This all makes life difficult for the knowledgeable client but impossible for the one who only builds 'once in thirty years'. Most major repeat business clients are able to keep themselves in touch with the complexities of procuring buildings suitable for their business. They appoint competent professional teams and agree appropriate procurement routes. For the 'one-off' lonely client the risk is high. The professional practices and research establishments and associations publish guidance on the appointment of the design team and alternative contractual routes. It is the difficulty of getting this advice to the 'one off', and that client organization's lack of knowledge of what is available, that increases risk in what is a large percentage of the construction market.

Much of the guidance documentation that the lonely client may be directed towards or come across by chance will be from the professional institutions of the architects, quantity surveyor(QS) and Association of Consulting Engineers (ACE) (the last representing civil, structural and building services engineers). Naturally this guidance will be biased, particularly with respect to the architect and QS saying 'Appoint me and I'll do the rest.' Alternatively, the single source responsibility approach may have appealed to the lonely client and he may have appointed a project manager or a chartered builder on a 'design and construct' basis.

2.2.2 THE PARTIES

What this implies for the site manager is the need to acquire knowledge of the background of how the parties came together in order to understand whether the client is 'knowledgeable' or 'lonely'. Does the lead come from the architect with subconsultants for structure and services? Is the project manager from a financial, quantity surveying base or has he in depth knowledge of construction management? If the site manager is to succeed he must have a clear understanding of the strengths and weaknesses of the client and the design team.

Depending on the form of contract the site manager's key relationship will be with the client or employer's representative or the contract administrator. This representation may be by the client direct, a project manager (PM) or the design team leader. The latter is usually the architect.

The client

Whether it is the ultimate client or his representative, the architect or project manager who is involved on site does not matter. What matters to the site manager is the 'client's' ability to be a good decision maker so that the work, or changes to it, may be planned and abortive work avoided. There is no doubt that the best jobs are those where the client is closely involved and makes speedy, well judged decisions.

The project manager

The risk the PM may bring to the site manager's door is an inability to obtain client decisions and operate with limited delegated powers. If the PM is the arm of a quantity surveying practice than there is risk, particularly from the smaller practices. They may be excellent at controlling the finances, but even this may go wrong as a result of not understanding the programming, construction and commissioning of building services. Truly independent project management firms or those operating as a business unit of a construction group generally have the necessary broad based skills that the site manager will recognize.

The architect

As well over 50% of building work is still let on traditional forms of contract the architect is most likely to be the employer's representative and designated as contract administrator. The risk to the site manager's project may here be from lack of experience, depth of resource and conflict of interest. In traditional forms of contract which often have a number of Provisional Sums and Prime Cost items, it is important for the site manager to constantly review the need for their removal and replacement by hard information. The author's experience with provisional and prime cost sums is that their impact upon planning and programming the work and its commissioning is not understood until too late. Examples of this are to be found in the sums allocated for decorative light fittings, kitchen and process equipment.

The multidisciplinary design team

A multidisciplinary design team does not always attend site meetings with the necessary level of representation on building services, attempting to cover these with the architect and/or structural engineer.

Single practice disciplines

Whether appointed by the client, the PM, or in the case of structure and building services, as subconsultants to the architect, the site manager will need to know the links of allegiance. Too often subconsultants are overdeferential to the architect. This can lead to situations where BS consultants crucially delay matters of concern in their specialist area, in the expectation that they will be brought forward by the architect. Be it a multidisciplinary or separate discipline design team the site manager should be constantly aware of the ebb and flow of harmony among the DT members. By being proactive and understanding the scope, content and phasing of the project's building services the site manager can also ensure that the DT turn in an adequate performance.

2.2.3 TERMS OF ENGAGEMENT

It is imperative that the site manager understands the terms upon which the BS designer has been engaged. The ACE schedules of duties previously known as 'Performance', 'Abridged' and 'Full' have now been replaced (see section 1.5.2, 'Consultants'), by Agreements A(2), B(2) and C(2) summarized below:

- Agreement A(2) – For use where a consulting engineer is engaged as lead consultant;
- Agreement B(2) – for use where a consulting engineer is engaged directly by the client, but not as lead consultant;
- Agreement C(2) – for use where a consulting engineer is engaged to provide design services for a design and construct contractor.

Each ACE Agreement [1] comprises *inter alia* the following:

- Memorandum of Agreement
- Annex to Memorandum – requirements for collateral warranties
- Conditions of engagement
- Appendix I – services of the consulting engineer
- Appendix II – remuneration of the consulting engineer.

The list of building services work elements in Appendix I above has been correlated with those of the CAWS and included in this book as Appendix B.

Agreement B(2) Appendix I – services of the consulting engineer under 'normal services' describe in clause 2 the service that most closely parallels the old 'Abridged' duties, which are usually discharged through the provision of a specification and general arrangement drawings. Clause 3 'Additional services (not included in normal services)' provides the listing of duties that could take the service up to the equivalent of the previously designated 'Full duties'.

Of all the alternative ranges of duties that a designer may be appointed for, those defined as 'Performance' carry the most risk to the contract. To discharge performance duties well demands the application of the design organization's 'best brains'. The designer must be very experienced to be capable of visualizing all the services systems in his mind in order to describe the requirements for design development well, with the minimum of documentation.

It is only within Agreement C(2) that any listing akin to the old 'Performance' duties can be found. In agreement C(2) where the consulting engineer is providing the service to a design and construct contract, this agreement will in many cases only be signed up by the successful contractor and designer. Therefore prior to contract award some other simpler short form, or two stage service agreement may apply with appropriate listing of duties. Such a listing of duties is suggested in Appendix C.

The duties for tender period would be appropriate for the first stage, with those for the design, installation and commissioning to handover phases forming the second stage appointment. These second stage duties would also be useful in reconciling any differences in the duties of the novated consultants.

2.2.4 DIVISION OF RESPONSIBILITIES

The Latham Report [2] '*Constructing the Team*' quotes one aspect of the design process crucial for the success of a project as being 'the co-ordination of the Consultants, including an interlocking matrix of their appointment documents which should also have a clear relationship with the construction contract'. The site manager should remind his contracts manager/director of the need of sight of this document and note any shortfall perceived in the service, for example, what BS elements place a design responsibility upon the installing contractor or any of his subtraders (see also clause 1.5.5, 'The installer').

Most BS designs will have been tendered on details that are not fully coordinated, this being left to the installers. The BS designs are therefore very often not much more than numerical solutions waiting to be proven through the working drawings, fit at the construction face, and evaluated post-commissioning. The potential for risk is considerable. In order to mitigate this risk, particularly at the 'fuzzy edges' of interpretation of consultants' and installers' responsibilities, the BSRIA led the production of their Technical Note TN8/94 *The Allocation of Design Responsibilities for Building Engineering Services – a code of conduct to avoid conflict* [3]. As a steering committee member and chairman of the

Table 2.2 Allocation of design duties: example of general design activities for community centre and sports hall (Based on BSRIA TN 8/94.)

	Design activity	Allocation of responsibility		
		Designer	Installer	Other
1.1	**Production of drawings**			
	Sketch drawings		As necessary	
	Schematic drawings	D		
	Detailed design drawings	D		
	Co-ordination drawings	D*		
	Installation drawings		I	
	Installation wiring diagrams		I	Controls/panels
	Shop drawings		I	Subtraders
	Manufacturer's drawings		I	
	Manufacturer's certified drawings		I	
	Record drawings		I	
	Builders' work drawings		I	
	Specialist drawings		I	As necessary
1.2	**Spatial coordination (i.e. overall responsibility for resolving difficult spatial clashes)**			
	*Bestdes plc drawings issued for tender show: Coordination by segregation of services along the primary and secondary horizontal and vertical distribution routes.			
	Generally single line plant room layouts, some with sections and elevations.			
	A double line arrangement of the multi service HVAC/PHE duct behind the central core toildes.			
	Proposed location of electrical/voice, vision and data services in duct behind main lift shaft.			
	These drawings have been prepared by Bestdes plc to confirm that the building services designs are capable of finalization of coordination by the installing contractor			
1.3	**Confirmation of plant or system sizing**			
	The designer is responsible for all installed plant and system sizes/capacities other than for those items which are indentified below. These items require final confirmation by the party indicated.	D		
	Plant items/systems: (to be completed by designer)			
	Fans and pumps as scheduled in the specification		I	

Table 2.3 Allocation of design duties: selection of plant and equipment for community centre and sports hall (Based on BSRIA TN 8/94.)

	Design activity	Allocation of responsibility		
		Designer	Installer	Other
2.1	Review the client brief and identify those client priorities which will influence the choice of plant. This may include consideration of factors such as initial cost, life expectancy, reliability, maintainability and environmental impact	D		
2.2	Identify project limitations which may influence the choice of plant. This may include consideration of factors such as space and weight limitations and the need to comply with health and safety legislation	D		
2.3	Where appropriate, consider the possible application of packaged system solutions, i.e. plant and systems designed and supplied as a package	D		
2.4	Prepare a description of the main performance requirements of plant items. This will involve establishing provisional values for the nominal capacities of plant, the range of operating duties anticipated, diversities applicable on maximum calculated loads, and the requirements for standby capacity	D		
2.5	Prepare descriptions of essential design features for plant items. This may include providing details of the expected quality of construction and finishes, any essential energy saving features, the acoustic performance, the availability of spares and the compatibility of the plant with other equipment	–		
2.6	Select on a provisional basis those manufacturers' products which most closely meet the project requirements of performance, quality and budget as established from previous duties 2.1–2.4	–		
2.7	Evaluate the impact of provisional plant selections on the overall building design. Advise on the need to amend the building layouts or structural details accordingly. Confirm provisional plant selections	–		
2.8	Advice on the need for pre-selection of plant. Where appropriate, invite quotations, report upon offers received and select equipment	D		

Table 2.3 *Continued*

	Design activity	Allocation of responsibility		
		Designer	Installer	Other
2.9	Incorporate ~~provisional and~~ pre-selected plant makes, models and duties in the specification ~~In the case of provisional selections include the~~ names of alternative products which comply ~~with the selection criteria~~.		I	
2.10	Prepare a report in consideration of any alternative plant selections proposed subsequent to the issue of the tender documents. Advise whether the alternative complies with the selection criteria established from duties 2.1–2.4	–		
2.11	Advise whether the alternative suggested is acceptable	–		
2.12	Fully re-evaluate all parts of the services and buildings design which may be affected by acceptance of alternative plant selections	–		
2.13	If accepted, amend the design to incorporate the alternative item of plant	–		

commissioning activities subcommittee the author contributed to this document. With its origins in the 1991 Futures Workshop Adversarial Attitudes Syndicate, to which the author was a delegate, this technical note came from industry wide cooperation involving the ACE, the Chartered Institution of Building Services Engineers (CIBSE) and the Heating and Ventilating Contractors Association (HVCA). Chaired by BSRIA with representation also from major contractors, PMs, building services consultants, designers and installers they identified the problems and proposed the solutions in the form of responsibility allocation schedules. Tables 2.2 and 2.3 show completed examples of pro formas 1, 'General design activities' and 2, 'Selection of plant and equipment for a community centre and sports hall'. Note in this example the designer has preselected plant makes, models and duties for activity 2.9, crossing out the second sentence. If the tendering installer were to submit a compliant tender, but in addition provide alternative plant selections, this could involve the designer in preparing a report against activity 2.10, the evaluation of which would involve the client paying the designer a further fee for this additional duty outside the scope of the designer's terms of engagement.

It is hoped that the allocation of design responsibilities for building engineering services will be encouraged in the following ways:

1. By knowledgeable clients when calling for fee bids against defined ACE terms of engagement, also requesting completion of the responsibility schedules. This should enable like for like bid comparisons.
2. By consultants and included in tender enquiry documents for building services contractors. This will minimize conflict through misinterpretation.

Whether or not the responsibility schedules have been used in the above way there is no reason why the site manager should not ask the design team (BS consultant) to complete the schedules. The contractor should then ensure he has a clear understanding of their compatibility with the services tender to be awarded. The services contractors agreement to the scheduled interpretations can then be sought at the pre-award meeting. If it is presented to the services contractor after the award of the contract this may create difficulties; but these should at least be resolved before start on site.

No apology is made for the extensive treatment of the need to remind site managers of the variety of design duties of which they require knowledge. Suffice to quote from the Latham Report [2]: 'It is vital that the Contractor knows who is responsible for which elements of the design and when they will be available.'

2.3 The impact of the chosen contractual route

2.3.1 GENERAL

The chosen route is very rarely best for the building services technology, but for the project overall. Contractual routes do seem to have something of a cyclical nature about them. After a run between the mid-1960s to the mid-1970s there has been a resurgence of the design and construct contract for small to medium size projects. These run in the shadow of project management, the other single point responsibility format. They have gained in popularity along with construction management from the demise of the management contract prevalent throughout the 1980s. These contractual arrangements bring risk in varying ways and degrees for the site manager. While the contractual route should always be that which is in the client's best interests, regrettably this has not always been the case. Knowledgeable clients have always sought to improve on the current choices. Despite the odd disappointment, looking back over time most clients will feel that their present routes are currently best for them. Some routes determined by QSs and architects have been made from a limited knowledge base and thereby serve the practice rather than the client. Figure 2.1 shows the apportionment of client/contractor risk for the common contractual routes.

Whatever the route the chances of achieving success are greatest, and risk to the site manager's company the least, on contracts where the

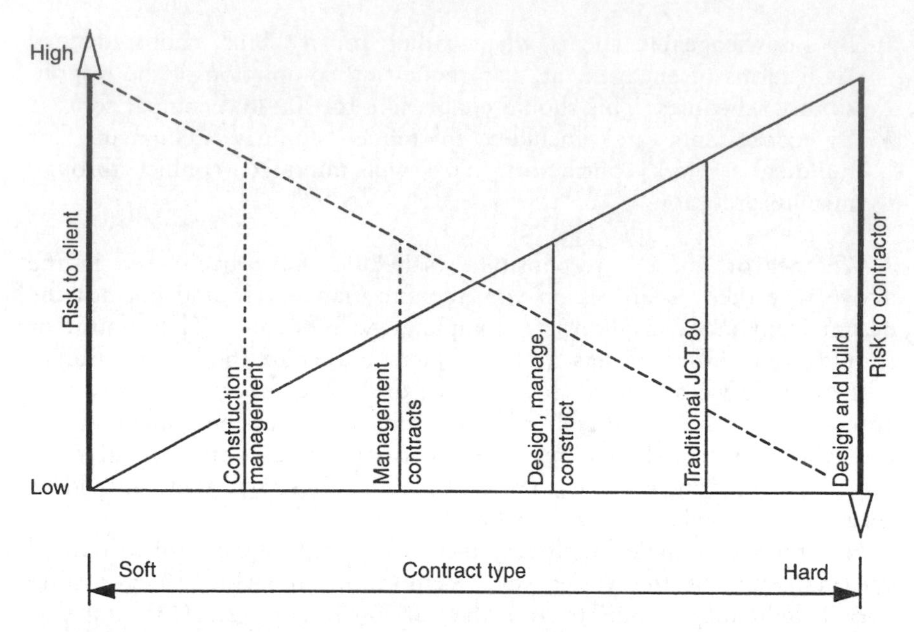

Figure 2.1 The apportionment of risk between client and contractor for the common contractual routes.

client is his own representative. These provide the site manager with the shortest communication line, and should ensure there are no ambiguities through messages degraded at the interface of many links. Short links are strong. The client will demand 'Why?', and the site agent can satisfactorily demonstrate with greatest effect, in real time, at the work face.

Clause 2.3.2, 'Traditional and management contracts'

At the beginning of the 1980s traditional standard form JCT 80 contracts were termed 'hard' and compared with 'soft' management contracts (MCs). By the end of that decade as the industry collapsed into recession the fee MC had become as hard and risky as the traditional. The reason for the degradation of the MC had a number of contributing factors. Many major contractors seized the opportunity of softer contracts and set up management business units. Often these units were inadequately resourced by staff who had only experienced working on traditional 'adversarial contracts'. They brought the wrong attitudes and felt exposed by the essential openness of managed contracts. Secondly, the QS or cost consultant also steeped in experience of traditional contracts had to learn the principles of management contracting without a standard contract as a guide. This scene was also new ground for design professionals. They had to learn how to define the scope of a package of work in sequential order, e.g. groundworks, frame and cladding and prepare the relevant tendering information of specification and drawings, without the benefit of understanding the full impact that elements such as false floors and ceilings, partitions and building services would have. This failure to understand the need for enabling information and inability to visualize the effect of yet to be designed elements caused many problems. In short, the skills were not there. As claims flooded in from package contractors and the nature of later elements of work became defined, the management contractors requested reimbursement for the assignment of more staff resources. Many clients could do no more than watch helplessly as the theory of having a professional team of which the builder was part, closely controlling all aspects of time, cost and quality, disintegrated behind a weekly supported facade of monthly reports. All parties learned the new politics quickly and found these had much in common with hard traditional contracts. Thirdly, the client's contractual adviser, usually QS cost consultant, and on some management projects designated the PM in order to wrest not only control of the cost plan but of the fee, determined the latter should be fixed. As fee – read profit – was earned by the management contractor in relation to the deployed staff resources, a fixed fee determined staff levels – not the best way to run a job yet to be designed. By the time the standard management contract came along in 1987, the contract advisers had decided that staffing levels proposed by management contractors at contract award would be fixed. Many projects were in the situation where fee and staffing costs were fixed irrespective of how the latter might be affected by the ultimate number of packages of work to be managed or the work content. Every management contractor has some horror story to tell of this period of having bid a 20 package contract with 10 staff, ending up with 40 or more packages and 25 plus staff. But still the management contractors bid for the jobs. These were often major projects bringing important cash flows through the books.

If this appears to be a tirade against management contracting, the appearance is deceptive. Many jobs were completed to time, cost and quality. But the reader is reminded that this section is about risk, and contract preferences run in cycles. Management contracting is still a very viable form but the site manager should be aware of its difficulties for it can suffer from the same painful compression of the BS elements, particularly their commissioning, that is prevalent in traditional contracts.

Whether because of designers seeking to pass on design risk or a desire to embrace the expertise available in the industry, many traditional JCT 80 contracts are 'with Contractor's Design Portion'. Sometimes as much as two-thirds in value and number of elements may have their design concepts novated but development controlled by selection from lists of preferred contractors and suppliers. For each of these contractor's design portions there will be a contractual requirement for the subcontractor or supplier to enter into a design warranty with the client/head employer. A line of direct access between employer and subcontractor or supplier, in the event of design failure, is not the route of contractual payment. The builder is protected from design liability but embroiled in any ensuing action, for which the recovery of costs may be difficult.

Benefits, and there have been many, have flowed from management contracting into traditional contracts. The pre-qualification of contractors, placing emphasis on the team led by a proactive site manager with a high level of communications skill, is good practice. There is no reason why this should not be carried further down into even smaller value contracts. The calling up of quality and safety plans, method statements, activity sequences and detailed programmes find their best practice origins in the managed contract.

2.3.3 CONSTRUCTION MANAGEMENT

Perhaps the removal of cash flow from contracting groups with construction management (CM) business units will help to distance this form of contract from too close an association with management contracting. This has yet to be seen. Certainly the best practice adviser cum agency role of the construction manager has potential for greater acceptability by designers, for influencing trade packaging and associated design information for tender. As a contractual arrangement CM is 'softer' than managing contracting. Much of the ability of construction companies to carry out this and any other form of fee building depends on how they are perceived by the client and the professional team.

Be it an independent CM firm or arm of a contracting group, the function of the site manager, or whatever project title that function is assigned, is still in charge of the site. When the Health and Safety Inspector visits the site it is the CM firm's site manager who holds up his hand in response to the question, 'Who is in charge at this site?' This should cause less of a problem under CM than perhaps any other contractual arrangement. Although, in order to avoid a conflict of interests, the CM's site manager should not have been appointed as the planning supervisor (PS) under the Construction Design and Management (CDM) Regulations he will have been closely involved with the latter before taking possession of the site. This involvement will arm the site manager with the health and safety issues considered by the building services designers. From the promulgated staff health and safety plan for the finished building will come knowledge of risks that will be the site manager's responsibility during commissioning and handover. The site manager is reminded that all of the site's health and safety policies will pass to the PS for inclusion in the project's health and safety file. This file will form part of project handover documentation.

2.3.4 PROJECT MANAGEMENT

The PM may be appointed with sole responsibility for procuring the design, construction, commissioning and handover of a project for which the choice of contract is for the PM. More usually project management is carried out either by the 'in house' team of a knowledgeable client who regularly procures buildings, or by the appointment of an independent firm of PMs. Sometimes these firms are autonomous units within construction groups. In the commonest arrangement the PM will represent and act on behalf of the client. In this executive role and purest form the PM would be responsible for the selection of the professional team and negotiation of their terms of engagements and determine the most suitable form of contracts through which contractors and subcontractors will be appointed.

Project management is therefore not necessarily a construction contract, but a method of overall project control. Within project management many forms of construction contracts may be employed; but most usually these veer towards the traditional. For the site manager whose firm has contracted to construct 'the works' less risk should emanate from design deficiencies, be they technical or informational. There is a caveat to this: that the PM has been appointed before the professional team. If not, the lately appointed PM may be faced with the impossible task of changing decisions that may bring difficulties for the construction site manager. An example of such difficulty is the commitment of funds to achieving a project programme proven to be impossible by the PM but put to the tendering contractors on the basis of 'let's see what industry comes up with'. Depending on the nature of the project and the tendering contractor's in house expertise for appraising building services the project is likely to proceed with a winning contractor who has committed himself in ignorance. Up at the interview table prior to award the contractor will answer all questions concerning programme in the way the PM team would wish to hear them. Good advise from the second and third lowest tenderers will languish in the waste-bin, its pleas for 'discussion on matters of programme' attenuated by the euphoria of a tender inside the cost plan. The winning contractor striving to achieve an impossible programme will certainly degrade the quality of the product. Be certain that this shortfall in quality will manifest itself at the interface of services and the building elements particularly finishes, and the cutting of corners at commissioning.

2.3.5 DESIGN AND BUILD, AND DESIGN, MANAGE AND CONSTRUCT

Was the strong cycle of the revival of these forms of contract in the 1990s a reaction to the perceptions of costly layers of management and client difficulties in achieving financial redress for performance failure? Or was it lessons learned from the earlier D & B cycle that led to the present innovative enquiries?

D & B contracts place the greatest risk with the contractor, for he is responsible for the design. Nowhere is that risk greater than with building services. In the best organized D & B contractors the site manager may have lived from enquiry, through tender design up to contract award. If fortunate enough to have done so he will have acquired an understanding of the risks devolving from the tightness of the brief and selection of BS systems. Knowledge will have been gained of the strengths and weaknesses of 'in house or brought in' design services.

Analysis of current D & B enquiries shows that over 80% of them come with various elements of pre-design. These not only, but mainly, related to BS, come with conditions requiring confirmation that the numerical criteria of the brief will be met, particularly concerning the internal climate, noise and lighting.

Where BS systems have been selected in principle if not in detail, then the tenderer's confirmation that their suitability and space allocation is accepted is sought quickly. A variant woven into this format is the request or invitation for the concept designs to be novated for development by the winning contractor. The contractor that markets in the D & B field without in house building services expertise is courting grave risk. If the site manager is assigned to the D & B project only at the time of award he would be wise to seek assurances that the BS criteria can be achieved by the selected building services systems. With time the greatest difficulty that the tendering services contractor has to surmount, designs are most likely to have been developed only in order to assess costs. These assessment/estimates are easier to calculate for the building elements than they are for the services. With short tendering periods very few services schemes are fully designed in the not untypical three- to six-week bid period. Appraisal of the employer's requirements, development of team strategy, use of in house or external services consultant to customize the building services brief, leaves say two to four weeks for the M & E contractor, who will go through the same appraisal. On a multi-service enquiry, documents will be copied to the four corners of the office – mechanical, electrical, public health and fire (e.g. sprinklers). Calculations to determine number and size of major plant and equipment are prepared for boilers, chillers, generators, switchboards, water storage tanks, pumps and calorifiers, etc. Telephone enquiries will bring a mix of budget/standard item prices. Schemes will be laid out (sort the coordination clashes out later), the quantities taken off and fittings (bends, tees, junction boxes, etc.) are included in the 'run'. Is there time to get prices in for the sublets (ducting, insulation, controls – building management systems (BMS), fire detection and security)? What about the contractor's request to put forward alternatives? What about displacement ventilation and chilled ceilings? Okay, do it on a cost per square metre?

The chief estimator receives four BS estimates containing varying degrees of risk. The tender meeting is 5.00 p.m. today. The D & B contractor receives a fax of the BS tender form with a single lump sum, but no supporting breakdown. A flurry of faxes provides the breakdown. A quick comparison across the computerized shreadsheet shows up the tenderer's highs and lows. Explanation is sought by fax and phone. The lowest services price is included with the second lowest alternative. 'We still haven't got the running costs or sample of maintenance contract?' Hey! I've just discovered the lowest alternative we've included requires us to put in double glazing with low emissivity glass – what's that?' Somehow it all comes together and is remarkably well presented. The interview goes well, your services manager was well prepared. You have got the job, but it's riddled with risk, has to be designed properly, and starts on site in four weeks time. 'Where's the drainage layout?' 'Don't forget the earth rod pits!' 'Can't they bring the gas main in at the weekend? Otherwise I'll be without site access for twenty-four hours'.

Enough author's self indulgence, let us return to sanity and consider design, manage and construct. This is a single point responsibility but one in which the client usually involves himself in the contractor's appointment of designers and package contractors. These may be reduced in number, packaging the work under a number of prime contracts. It is usual for the construction cash flow to remain with the contractor with design fees paid by the client. The contractor will control his managing staffing levels. Profit will be the fee. This is perhaps, the most suitable contractual arrangement for giving a client a guaranteed maximum price (GMP). The contractor may not know the moment when the client is going to ask for the contract to be converted into a GMP. The response must be very carefully considered. If earlier in the contract the risk from the BS angle is less. Before responding the design can be audited and reassurance sought that it will perform as required. A GMP requested once commissioning of BS has commenced is of quite different order. Time is short and the client impatient at the delay while a belated design audit is carried out. Meanwhile commissioning is progressing apace and may be showing up some problems. At the beginning of the job the contractor has all the contract value to spend and can buy in performance assurance and include it in the GMP. At the end of the project with most of the value expended and perhaps only 10% of the project value to come, the contractor has difficulty in asking for a further 10% to cover the risk of services failing in some aspect or other. The site manager who has stayed closely involved with engineering services has an important part to play in discussions with the contracts manager/commercial director in capping the contract value.

2.3.6 SUMMARY

In ascending order, risks arising from the BS element and the chosen contractual routes are summarized as:

- Traditional including those newly derived from the national engineering contract (NEC): lowest risk if fully designed and installed by a well procured subcontractor. Rising in risk according to the amount of undesigned Provisional Sums and Prime Cost items.
- Traditional with contractor's design portions: risk arising where the services elements form part of the design portion, or interface with other design portions such as cladding. In this instance difficulties have been known to occur with fixings for perimeter heating, HVAC equipment and permissible levels of air leakage.
- Management contracting: incompatible packaging of work giving rise to difficult subcontract management.
- Construction management: as for management contracting plus greater risk due to the monitoring-only role of the construction manager of the on-site works. To a degree this is compensated for by a better quality of input to the design appraisal, constructability and commissionability of the building services.
- Design and build: clients see this as the route with least risk. The D & B contractor sees it as a business of higher risk with commensurate reward. A number of D & B contractors provide an excellent product, but many have a common weakness and expose themselves to unnecessary risk through an inadequate comprehension of building services. This risk manifests itself in a number of ways – the lack of in house building services expertise, and the employment of cheap, lower division, underresourced consultants with narrow experience. Others without design expertise, and recognizing the pressures of tendering time, go to D & B BS contractors. Regrettably many previously excellent BS firms have been forced to disband their design departments or operate a shell, buying back agency staff on an as needed basis. A glossy presentation behind a subeconomic price can, by its attractiveness, deter an unsuspecting client from probing the disguise into clutching the single point responsibility. After all, this principle with speed, low cost and acceptable quality was the reason for the D & B route.

Project management does not feature in the above contractual order of risk, being a method for managing a project for which selection of the building contract route is but one function. Despite finding the addition of another layer of management irksome the building site manager is to be encouraged to find out how satisfied the PM is with the building services design. Questions along the following lines should be asked:

- Is this proven technology?
- Are there any aspects that make it difficult on this project?
- What is innovative about it?
- Have the innovative features been tested?
- Were the results satisfactory?
- Can we have copies of the reports?
- What aspects of design most concern you?

The best PM will be happy to inform the building site manager.

The site manager is reminded that this is the expected order of risk arising from building services under the contractual arrangements he may experience. There is little the site manager can do to mitigate the impact of the chosen route if the site manager is not part of that process of choice. Frustrating though it is to be embroiled in difficulties not of your own making there is much that can be done to assess areas of known and unknown but potential risk. Fore-armed with this knowledge the site manager will be able to make a contribution to the resolution of difficulties and remain in control of his site. Without this knowledge of contractual route risk the site manager will become no more than the figurative head as the project becomes swamped with the opinions of conflicting expertise talking an alien language.

2.4 Understanding technical risk

2.4.1 GENERAL

The technical risk of BS is more easily understood from a platform of basic knowledge. The most basic framework is provided in Chapter 1. Variously site managers will acquire deeper knowledge through training, including CPD, and will be able to flesh out the framework with their understanding of specific types of systems. All those who attain the status of site manager will have some system-specific experience.

2.4.2 THE WAYS AND MEANS OF SYSTEM FAILURE

Building services systems may fail in any number of ways:

1. by component design
2. by manufacturing defect

3. by installation defect
4. at element interface
5. by performance (incorrect system or component selection).

These failures are no different from the failure of other building elements. It is at the moment of discovery of failure that BS differ. This requires further explanation. Beams deflect, doors and windows open and whole buildings may settle. The dynamics of other elements of

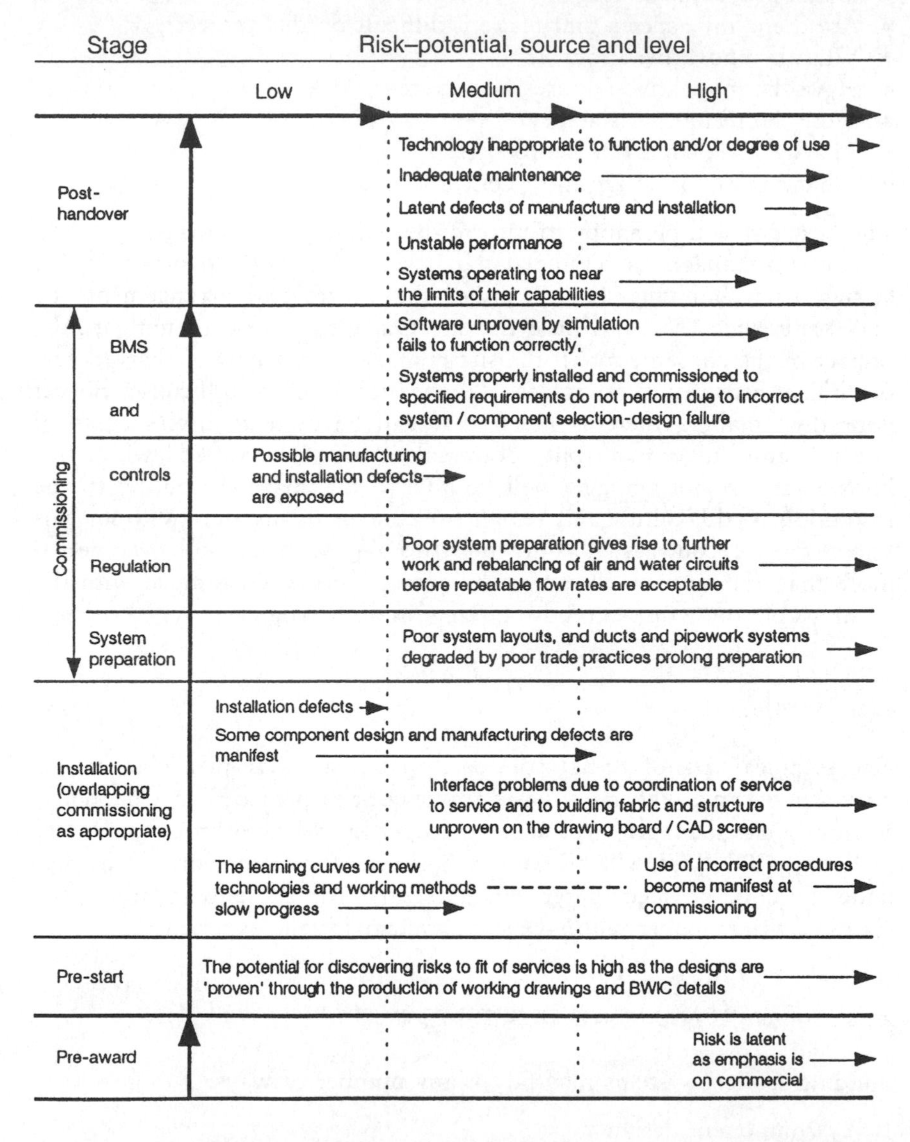

Figure 2.2 The discovery of risk in relation to building services project stages.

buildings are very limited. It is the dynamic nature of BS and the need for them to be tested, proven and witnessed to the satisfaction of designers, authorities, utilities providers and insurers that affects the moment when a failure is discovered. As systems are built to the stage of their static completion, quality control and corrective action will manage the rectification of some types of failure associated with (1), (2) and (3) and possibly some of (4). Included here are the construction tests of liquid, air and gas pressure testing, for leakage. In the main these are the identification and rectification of visual failures of compliance. They are dealt with progressively along the way. A clear point when failures manifest themselves is at the time of preparing the water and air conveying HVAC systems prior to commissioning. Those apparently well constructed pressure test proven pipework systems may have, in addition to the millscale of manufacture, excessive welding slag, hemp and paste and other detritus in them due to leaving open ends. These unseen defects can be managed away over time, and herein lies the risk to the site manager's programme.

Even with a good standard of system preparation, commissioning can give rise to another set of problems. If design is suspect, ensuing corrective variation may have a knock-on effect in delaying other trades. The witnessing of system tests is now upon us. Smoke management tests can bring out a problem at the interface of services and building elements. With the right quantity of air pumped into them staircases and lift lobbies won't register the necessary pressure differentials. Although 'approved' the installed fire alarm system requires the installation of additional sounders. HVAC performance tests may show an inability of the installed systems to maintain the specified internal climate. The discovery of risk in relation to building services project stages is shown in Fig. 2.2.

Occasionally a project may be doubly unfortunate and the site manager burdened by a poor design installed by a bad contractor. Larger projects may exhibit some of the failures, but few are spectacular. Large jobs benefit from better 'premier' division designers and better managed sites, all contributing to the availability of a greater depth of expertise when things do go wrong.

2.4.3 APPRAISING PROJECT RISK FROM BUILDING SERVICES DESIGN

The site manager should not feel defenceless to the risk from the BS on his project. For the same reason that he will have armed himself with the nature of risk from the type of contract over which he had no choice, the deeper the knowledge of BS the greater can be his influence. Irrespective of the contractual arrangement Appendix D is a matrix score sheet identifying the potential pitfalls arising from the design of building services. From this it will be observed that projects with a high value of building services which include air conditioning within an overall project value of £2.5–10 million have a high risk. This risk is increased if the air conditioning criteria call for close tolerance performance with respect to noise, temperature, and particularly humidity. The site manager should have a first shot at assessing his project against the matrix. It is strongly recommended that a second opinion be sought from an experienced BS engineer. This should provide further confidence to the site manager and refine the risks for which a strategy can be evolved.

2.4.4 GETTING AND USING EXPERT HELP

If from the first appraisal of the matrix the manager has defined a high risk content he would be wise to seek expert help. Such help is best provided from within the site manager's own organization through the support of a BS manager (BSM). Over time, through association on a number projects, a valuable understanding can be established, but only if the BSM has the necessary breadth of experience and depth of knowledge of all building services.

A considerable number of major contractors now employ building services engineers in roles of varying status. The site manager is indeed fortunate if he works in an organization with those resources. Increasingly, contractors are realizing the importance of BS staff resourcing, but as yet are unable to provide this back up to every construction project. A considerable number of projects, even with values above £2.5 million, may find the site manager without any structured building services support from the in-house resource. Finding himself in this situation it is for the site manager to be proactive in mitigating risk from building services by requesting assistance. This is not to be seen as an inability to cope but the sign of a good manager, who from a simplistic technical appraisal, calls for a deeper verification. It is a very unwise contracts manager/director who fails to provide requested services support. The provision of such support may not have been envisaged when costing the staffing, but quantified expenditure on such support may save unquantifiable costs in the resolution of problems, as liquidated and ascertained damages hurtle towards the construction company.

For the smaller project, most likely to be on a traditional contract, the minimum provision of support should be by a senior BSM who should:

1. Visit site before the subcontractor commences and study the specification and drawings, both design and working, and summarize the technical risk in discussion with the site manager. It is important that this is a two-way dialogue. He should also comment on the services programme.

2. Attend the pre-start site meeting. With the site manager and services contractor(s) run through the pre-award meeting agreement (see section 3.1.3).
3. When advised by the site manager that examples of all trades work are available, visit and inspect. This should be not later than a few weeks after installation has commenced.
4. When advised of their availability, comment to the site manager on the subcontractor's proposed method statements for system preparation (e.g. flushing and cleaning pipework), and commissioning (regulation of systems).
5. When called upon to do so, support the site manager in any difficulties he may be experiencing with the subcontractor's inability to prove system performance to the designer.

The above is the absolute minimum support, and we are talking of six to ten days services input for simple services with a value of 15–20% of the overall cost, on jobs of no more than £2.5 million.

As projects go up to the scale of overall value and technical complexity the input of the services engineer should be managed to provide visiting support on a more regular basis. This may rise to two to three days per week. Above that period it is extremely difficult to split a services engineer's time four days to one. The most sensible approach is to provide the services engineer full time.

Certainly on jobs embracing both contractual and technical complexity, and these are most likely to be more major projects carried out on the MC/CM route, then the provision of a full-time BSM is essential. Since there is full-time services cover the site manager should keep in close touch with all aspects of BS. The experience gained will make him more valuable to his company and enhance his own career prospects. The confident management of BS and their attendant risks will see the industry provided with site managers competent in all building elements. As building fabric and structure become more closely integrated in the way they work to create internal climates, so site managers will hold the high ground in their industry.

Contractors wishing to employ BS staff in house or review the capability of their current resources may find the job description for a BSM included as Appendix E, helpful. The description is in a form that can be applied to traditional, D & B and fee managed contracts. The duties of services managers in project management organizations are shown in column 4.

2.5 Procuring the building services tender

2.5.1 GENERAL

As we move to the conclusion of this chapter on risk the emphasis is changing from the site manager's awareness, to what the construction company can do to properly support him. We have a good site manager with appropriately structured building services support; all that is needed now is the right BS contractor. It is hoped that by now an appreciation of the risks that can be attendant on building services will have created a force for change in their procurement. These elements that remain on the critical path of production for so long, and at the end of the project are tested, documented, and users trained in their operation, unlike any other building work, cannot and should not be procured through old established routines.

2.5.2 THE WRONG WAY

Commercial managers and contract buyers may shudder at the need for change, but time and time again contractors burden themselves with the wrong M & E firms. Tender lists are too long. Forced through slavish adherence to vendor databases created to service BS5750 registration, they fall foul of its shortcomings. Unless kept up to date, and this is costly, through the input of meaningful feedback, vendor databases quickly become dated. They are weakened by the speed of oscillation of the economic pendulum, affecting the managerial, technical and labour resourcing strategies firms have to employ to react to market conditions and stay financially viable. Enquiries are sent out by the uninformed, who wonder why the returns are incapable of analysis. Due to the high value of the services element the inclusion of the wrong services price through ignorance can be the difference between winning or losing a job. The worst scenario is to lose the job because too high a price has been included through fear of the unknown. The reverse also occurs through over keenness and the telephoned reassurance of a compliant bid. Most of what has been stated occurs on traditional contracts and to a lesser extent by D & B contractors trading without their own BS management resources. There is far less risk in the procurement of services contractors on fee format contracts. The outburst against procurement methods on traditional contracts is with the objective of properly procuring building services tenders. If this is seen as treating them as 'special' or 'elitist', so be it, but until changes are made and lessons learned from the best practice of D & B and fee management, then a large area of risk will remain with the industry. The post Latham Report [2] improvement in building contracts for the traditional procurement routes will only serve to expose even further, risks associated with procuring BS contractors.

2.5.3 A BETTER WAY

What follows is a base line methodology for building contractors in which they are encouraged to use the designated site manager and highest available level of their own building services staff:

1. Distribute a complete set of the project enquiry information to those assigned the duty of appraising the BS. How can they judge the inter relationship of building fabric and frame without the architect's and structural engineer's drawings? What hidden risks to delivering the technology are there in the contract conditions and specification preambles? See the chapters on commissioning and handover.
2. Appraise the building services using matrix Appendix D or, as a minimum, answer the key questions given in Table 2.4. A feeling for the degree of risk will emerge. A useful overview can be formed by studying Fig. 2.3, which is based on the analysis of volume business carried out under various contractual arrangements over 15 years. Table 1.5, giving building services costs as a percentage of overall job value, will be of assistance in risk profiling.
3. Check that all information is available for the tenderers to prepare their quotations. For example, even pre-designed jobs with sprinklers will usually require the installing contractor, who may be a specialist trader, to carry out the hydraulic design (pipe sizing) to the specified requirement of hazard classification. This information alone may be inadequate. Details of water pressure and integrity of supply will

Table 2.4 Initial appraisal of risk due to building services content of project

Key questions	Sample answers
Value of project	£8.7 million
Value of services	£2.5 million (27%)
Air conditioned	Yes
Building type	HQ offices
Type and capacity of services	Dense/extensive
	Humidity 50% RH+/–5%
	Fibre optic – structured cabling system
	Motorized light shelves
Designer capability	Small practice, unknown capability
LAD	50 000 pw or part thereof
BMS	Yes
Risk profile	High
If enquiry is for D & B, additional questions	
System types selected	Yes – displacement ventilation with chilled ceiling
Is design novated?	Yes – recommended concept design consultants are appointed by contractor

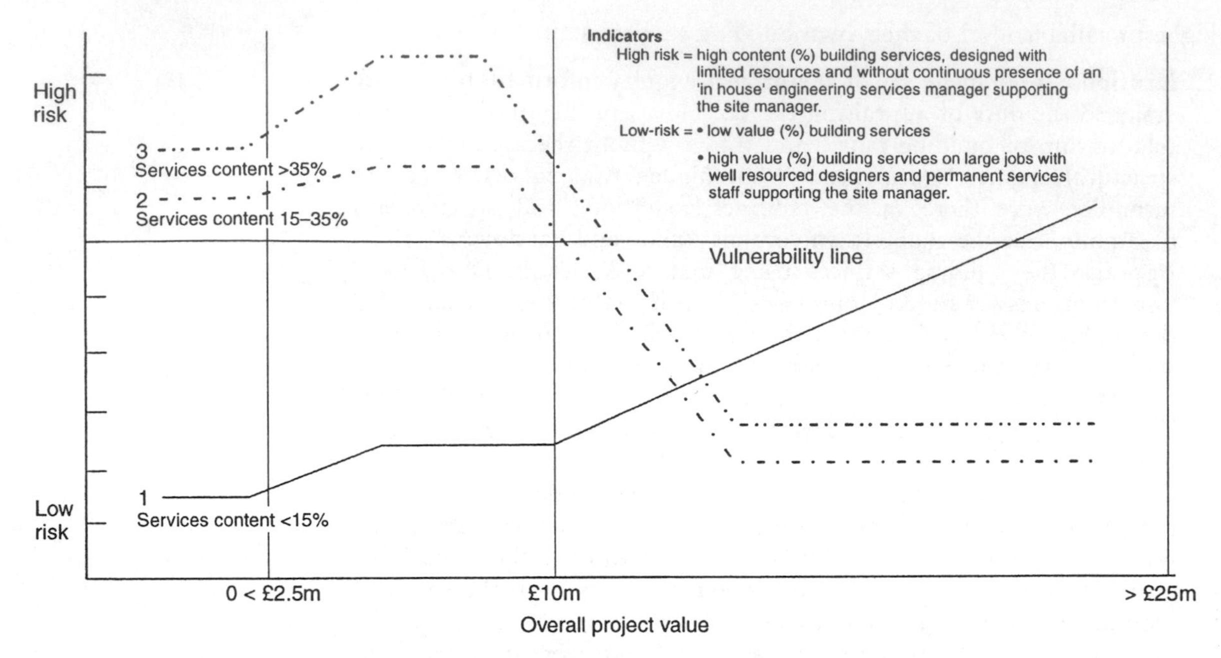

Figure 2.3 Profile of risk associated with building services.

also be required. This information should be obtained and sent to the tenderers.

4. Prepare a list of enquiry documentation. Table 2.5 is an example of tender enquiry documentation in the form of a production schedule. If a breakdown of price is not called for in tender invitation to the building contractor, it is recommended that one is used to aid the analysis of the building services tenders.

   Appendix F is included as an aid to preparation of a breakdown of tender. For a great number of projects the summary of headings on the first page of the appendix will be adequate. From the pages following the summary greater or lesser detail can be incorporated, recognizing the scope and content of the services in relation to the geography of the building and any phased handover requirements. This search for the most suitable breakdown specific to the project in question will unavoidably blur the CAWS alphanumeric listing of services families and system types.
5. Who is to tender the building services? While vendor databases may be the starting point of the commercial department, with knowledge gained from the technical appraisal unsuitable services contractors can be weeded out and only those considered suitable invited. Depending on the positioning of the pendulum in the swing between

Table 2.5 Production control of tender enquiry for building services with design development by installer

Project ______________________________

Tender enquiry documentation for: Building services

Item	Document	Draft	Word Processor Draft	Chkd	No. of Copies	Printed	Source: also used on previous job
1.	Enquiry Letter						
2.	Technical Brief (see Note)						
3.	Subcontract form						
4.	Extract from main contract conditions						
5.	Scope of builders' works and attendance						
6.	List of drawings issued with enquiry						
7.	(a) Form of tender (b) Breakdown of price						
8.	Preliminary schedule of builders' work requirements						
9.	Details of equipment and plant to be lifted by main contractor						
10.	Schedule of dayworks						
11.	Quality, environmental and H & S requirements						

Note: Technical Brief, Item 2, may comprise employer's requirements amended to include the following contractors needs: brief/scope; specification; list of approved suppliers; schedules of duties.

boom and bust the best advice will be to use the fewest number of contractors who, while knowing they are in competition, will retain some 'partnering affinity'.

6. The analysis of the returned tenders will involve removal of commercial and technical anomalies and present the site manager with further opportunity to understand potential areas of risk.

Having acquired a reasonable level of knowledge of the BS through the issue of a sensible enquiry and tender analysis the contractor will have mitigated a lot of the risk associated with the procurement of a building services contractor. The award of the overall contract will now allow the contractor to go forward and successfully manage with the subcontractor the installation, commissioning and handover of all 'the works'.

References

[1] The Association of Consulting Engineers, *Conditions of Engagement 1995.*
[2] Latham, Sir Michael (1994) *Constructing the Team,* HMSO, London.
[3] BSRIA (1994), Technical Note TN8/94 *The Allocation of Design Responsibilities for Building Services – a code of conduct to avoid conflict,* Bracknell.

Part Two

The Management of Building Services Contracts

Management strategy 3

3.1 Strategic management

3.1.1 PROGRAMME AND KEY POINTS

Figure 3.1 depicts a strategic management programme through which every BS contract must travel. This is followed by a schedule of key points in Table 3.1. These documents should be the site manager's home base – a constant source of reference at which he can remind himself at any point on his contract as to where his services subcontractors are and what their next objective should be.

3.1.2 DECLARING A MANAGEMENT STRATEGY

The practices recommended in each chapter of this second part can be applied immediately. However, overall project benefit can be achieved

Item

Duration of contract

Quality plan
Contract safety plan
Environmental plan
Working drawings
DWIO drawings
Method statements
Sample approval
Test and inspection plans (for construction)
Schedules
Electric/gas/water/drains connected
Construction (supervision and inspection)
Electric feeds to plant rooms
Motor control centres
Feeds to fans and pumps
Preparation of systems
Commissioning
Training instructions
Record drawings
O and M manuals

Handover

Watertight plantrooms

Legend:
Information flow
Continuous activity
Intermittent activity

Figure 3.1 Strategic management programme.

Table 3.1 The management of building services contractors: schedule of key points

Building Services contractors should work in a well managed environment in which the Site Manager shall ensure:

1. Work must not start before a quality plan or QMS Statement appropriate to the scope of the work has been approved
2. The contractor's environmental and safety plans are approved before work commences
3. The production of working drawings including coordination details shall be monitored through an agreed approval cycle
4. Builders' work drawings for bases, holes and fixings, etc. shall be monitored. The policy for holes shall be established to determine those to be formed, marked out and cut, and built e.g. in brick or blockwork
5. Method statements shall be required from contractors covering the basic construction of mechanical, electrical, public health engineering, lifts and escalators, sprinklers, hosereels and dry risers, etc. These basic statements shall cover pressure testing, non-destructive testing, earthing and bonding.
6. Method statements shall also be required for special services such as fire and security systems, embedded floor coils, diesel generators and uninterrupted power supply systems, etc., etc.
7. Contractors shall produce schedules of samples to be approved
8. Schedules are to be produced by contractors covering any specified off-site manufacture and procurement, including testing of items such as prefabricated work, switchboards, generators, chillers and air handling units (AHUs), etc.
9. Electricity, gas, water, drainage, telecoms and data utilities shall be programmed, installed, and enlivened in time for commissioning
10. Services construction shall take place in plant rooms when adequate weather tightness has been achieved and is capable of being maintained
11. Priority is given to installing the electrical supply to motor control centres/outstations so that feeds to fans and pumps are available to prepare systems for commissioning
12. According to specified requirements water systems shall be flushed and cleaned, and air systems blown through
13. Following pre-commissioning, that the commissioning process is to be taken through the logical stages of balancing, setting up controls, commissioning major plant such as generators, boilers and chillers
14. Specialists shall be employed as appropriate, to commission kitchens, UPS systems, data, telecoms and vision services, BMS controls and outstations
15. Fire detection and protection systems shall be commissioned to the approval of the fire authorities including any specified smoke pressurization systems. Particular attention shall be taken to ensure the specified building construction is capable of the appropriate air tightness
16. In addition to the general commissioning of the services installations, any specified tests and demonstrations required to validate plant/system capacity, environmental conditions, noise criteria, and modes of operation, e.g. normal, standby and emergency shall be planned, programmed and resourced
17. The specified training requirements shall be met through the preparation of programmes agreed with the employer/client/tenant
18. A programme for the production and approval of record drawings, and operation and maintenance manuals shall be prepared well in advance of handover, allowing at least two four-week approval cycles

Note: Do not expect a services contractor to willingly deliver the methods, management and control documentation covered above. It is almost certain the site manager will have to 'extract' performance from the building services contractor.

through the declaration of a management strategy for the BS contract. The employment of such a strategy could be in the form of a simple procedural statement. Naturally, decisions will need to be made as to when to declare the policy to the BS contractor.

The management strategy framework, for that is all it needs to be, can be formulated from the building blocks of Chapters 4 to 11. The blocks are of different shapes, sizes and materials, and for success they must be stabilized into a coherent policy. See Appendix G 'Declaration of management strategy requirements' for an example of a BS contract.

3.1.3 THE TIME TO DECLARE

There is an important choice to be made here. The builder will not wish, or need, to give an impression that compliance with his BS management strategy is onerous, to such an extent that tenderers increase their prices. The statement of management strategy to be complied with, if included in tender enquiry documents, should not be prefaced with words like, 'conditions' or 'special clauses for building services contractors'.

Those builders entering into joint ventures, partnering, or preferred contractor arrangements must look for a meeting of minds on their proposed management strategy which can be mutually adjusted to suit any particular project. Others may see the use of management strategy statements as only differing in detail from the range of subjects they normally discuss with a contractor at a pre-award meeting. If the requirements for complying with a BS management strategy are to be raised for the first time at a pre-award meeting it is most important that the details are sent to the BS contractor well before the meeting. Nobody likes to receive this sort of surprise, which although non-contractual, may at first sight create the wrong impression.

There is a great deal to discuss between the builder and the BS contractor at pre-award and it would be a good idea to have two separate meetings. The first meeting would cover the normal subject matter raised by the builder, common to any contractor, followed by a session on the BS management strategy requirements.

At the meeting to discuss the strategy the builder must set the tone and state that his objective is to receive confirming evidence from the BS contractor that construction, commissioning and the requirements for handover are proceeding as planned and meeting the specified standards. The documented assurance of progress should enable the builder and BS contractor to be mutually supportive when difficulties arise – as they will – for BS never fails to surprise in the rigidity with which it adheres to Donald A. Norman's version of Hofstadter's Law [1]:

- It always takes longer.
- It always costs more.
- It will always be harder.
- There will always be more.
- There will always be less than you expect.
- Even when you take into account Hofstadter's Law.

By the way, Hofstadter's Law says:

- It will always take longer.
- Even when you take into account Hofstadter's Law.

3.2 Relationships

Up until the arrival of the BS subcontractor, work has been carried out in construction stages: groundworks, foundations, frame, envelope, floors, cladding and possibly some brick and block work. This is work the site manager is most comfortable with. The organic growth to the specified geography is readily visible, starting with one or two subcontractors and manageable quantities of drawings and information. This work will almost certainly have included drainage under the building and some enabling works associated with services entries. Enabling works have allowed the pattern of construction stages to flow. Now, with the arrival of the BS contractor or contractors in combined or separate elements, mechanical, electrical and public health starts on site. The information, drawings, specifications, bills and schedules being referred to seem to quadruple and the pattern of work is not always obvious to the uninformed. It is this shift in workload that the site manager must recognize and be geared up to manage. What follows in fleshing out the strategic management programme and summary will cover probably 80% of project types and related services complexities and provide a working platform for the remaining 20% of even more complex and densely serviced projects.

The site manager's leadership and interpersonal skills can make the difference, even under the most difficult contract conditions, between a project in conflict or harmony. There will be problems for the site manager creating opportunities to use those skills and ensuring that all relevant parties involved understand their responsibilities for contributing to its resolution. In achieving problem resolution there will be some friction. It is the site manager who must ensure that the rotating speed of problems does not abrade them, beyond the hoped for highly polished finish, into senselessly burnt out relationships and on to the handover of a thinly disguised damage limited project.

Success in implementing the strategic management programme will depend on the quality of the site manager's leadership. Management is POCC – planning, organization, coordination and control. This definition will stand testing on any building project. Consider that a project must have a strategy and there must be a plan to meet the strategic needs. There must be organization to implement the plan. That organization is contracted to the builder, project and construction manager and they further subcontract the work to other organizations. All of these parties require to be coordinated, brought into proper relationship to combine and create a completed project. The activities of procurement and construction resourcing of plant, equipment, materials and labour must be controlled, as must the preparation, commissioning, testing, documentation, training and handover of BS systems. The site manager must be strong, and proactive on the basis of knowledge. How much knowledge? Sufficient to bring negotiating skills into play in the grey areas and firm up in the fuzzy edges of responsibility. Through the exercise of such skills problems can be resolved. The best site managers will recognize that the need for fair treatment of subcontractors can do much for project harmony and lighten the reins of essential control. The route for this harmony model must be through:

- the selection of suitable subcontractor(s);
- the provision of the attendances contracted for;
- working to agreed programmes;
- teams with appropriate organizational, technical and personal skills and the flexibility to create harmonious interfaces.

This model will not work without equable partnership in the objectives which must be reflected by the services contractor. Where this is not being provided as of right it is the site manager's duty to demonstrate inadequacies to the provider so that getting the correct performance does not degenerate into an acrimonious extractive process.

If there is one area above all others in which the site manager can do most to demonstrate his encouragement of the subcontractor it is to pay him fairly and promptly.

3.3 Pre-award meeting

Having selected the right subcontractors to bid, evaluated their offers, removed any technical and commercial anomalies and included them in the winning tender the contractor is now in a position to award a services subcontract. The site manager who has been involved up to this point will be far more comfortable at the pre-award meeting than the site manager who is told 'This is who you are getting'. Even then all is not lost. But some situation retrieval may be necessary for it is at the pre-award meeting that the die of future relationships is cast.

The purpose of the pre-award meeting depends on where you sit round the table and the industry's economic health, which may be anywhere between boom and bust. Wherever the industry is on the arc of that economic pendulum it is important to remind oneself that it is the norm – for the moment. However, the economic climate may affect corporate policy and the way in which it is discharged through the functions represented at the pre-award meeting; it is critical that the site manager remains aware that it is he and the services subcontractor who will be managing and constructing those elements of the project. For this reason there must be a consistency in the site manager's attitude to ensure that clearly defined responsibilities are established. It is worth looking at the scenario of a pre-award meeting taking place in times of economic difficulty. Depending on the size of the job the contractor may be represented by the:

- commercial manager
- contracts manager
- planning manager
- subcontract buyer

} office based

- site manager
- site surveyor

} site based

and the services subcontractor by:

- a director
- chief estimator
- contracts manager

} office based

- site manager – site based.

Consider the company and individual attitudes that will be brought to the table. Unfortunately the pre-award meeting has historically been the forum where final negotiations on price are mixed up with the builder extracting from the subcontractor promises of unspecified performance, and withdrawing his own attendances commitments to produce a situation the subcontractor did not envisage when compiling his tender. So we have on one side the builder looking for a lower price now that he has an order to place, and on the other side a subcontractor who has come for the order at the tendered price and also to talk about the job. It is in this potentially adversarial situation that the building site manager must position himself as a key player in 'constructing the team' (see the Latham Report). After all, it is this team who will be building the project.

The commercial members of the builder's team can do much damage by taking personal, and company, hard-line attitudes and springing on the unsuspecting subcontractor requests for further discounts. Tangible though these sums are, the request for them can produce in the subcontractor a change in attitude that leads him to seek recovery of reduced profits through claims, enhanced variation estimates and inflated dayworks charges. Counterclaims will follow from the builder and the battle lines are then drawn, requiring support forces that neither party has costed into the job. The design team is unwillingly drawn into this arena, while the client watches in dismay. Who dares – to ask for discounts at the pre award meeting – loses, and we have not yet fixed a length of pipe!

Although more prevalent in difficult economic times, the further discount – buying gain philosophy – is not uncommon during better trading conditions. Where and whenever it occurs the practice is to be abhorred and is not the route to the Latham objective to reduce the cost of building by 30% by the year 2000.

Dependent on the function, geographical layout and complexity of services, the builder will have decided in the preparation of his winning tender what in house building services management and supervisory expertise is provided. Whether this is to be full-time site support or visiting, it is strongly recommended that the most senior services manager/engineer to be involved in the project should attend the pre-award meeting. A pre-discussion on the meeting's objectives and the requirements should take place between the services engineer and the site manager.

An agenda for the pre-award meeting should be circulated to the attendees; a typical example is shown in the Table 3.2

Unless the subcontractor is aware of the scope and content of the meeting he is likely to attend with the wrong representatives. Certainly subcontractors should bring to the meeting the person who

Table 3.2 Agenda for pre-award meeting

1.0	Purpose of meeting
2.0	Tender
3.0	Organization
4.0	Quality/safety/environmental plans
5.0	Programme
	5.1 Delivery of information
	5.2 Construction
	5.3 Commissioning
	5.4 Documentation, instruction and handover
6.0	Meetings
7.0	Attendances
8.0	Protection
9.0	Site hours
10.0	Working rule agreements
11.0	Labour strength and welfare arrangements
12.0	Insurance
13.0	Valuations/variations/daywork
14.0	Correspondence
15.0	Order
16.0	Any other business

will be responsible for taking instructions and managing the site. On smaller jobs where the subcontractor intends to support it by a visiting manager or engineer on a one-, two- or three-day a week attendance, that person should also be present. It is with these members of the subcontract team that the building site manager and his services engineer must establish a clear understanding of requirements at the interface of the subcontract. This will be the foundation of their relationship throughout the subcontract. Minutes of the meeting should be circulated and appended as one of the subcontract documents attached to the order.

The worst situation is where there is no site management representation on either side. In these cases, agreements reached tend to concentrate on the commercial aspects, with the builder negotiating the provision of fewer attendances and the services contractor 'happy' to give these away providing he is able to return to his office with the order. Later, when the subcontractor starts on site neither party is aware of the pre-award meeting agreements. The subcontractor asks for attendances he considers 'normal' only to be told his firm has agreed to provide these themselves. The seeds of conflict sown at the pre-award meeting begin to germinate.

It is not intended to go through every item on the agenda in detail. They will find their natural position as the management of a BS contract unfolds in later sections. But some key management issues must be addressed.

3.3.1 OBJECT OF THE MEETING

It is most likely the value of the BS subcontract will represent in financial terms a large percentage of the project's overall value. The pre-award meeting presents the opportunity for the contractor to put in place a key player in his team. Of course this is subject to the common desire of both to play on a level pitch, on the right side, with the same ball. Setting the tone of the meeting is most important: 'subject to agreement on the agenda items it is our intention that at the end of the meeting we will be able to say that you will shortly receive our order for the works.' A statement of these intentions will do much to set a harmonious tone for the meeting which can be closed out with 'when we have exchanged confirmation on matters we have just discussed we look forward to placing our order with you'.

If the meeting has been difficult, but for whatever reason the contractor is committed to securing that particular building services subcontractor, then a relationship has been established that may jeopardize the success of the project. In constructing a team all parties must want success.

3.3.2 ORGANIZATION

On- and off-site organization is something which must be addressed at the pre-award meeting. Some indication of this may have been called for and responded to in the subcontractor's tender submission. Beware the standard organograms of rigid company structure and standard site management team. The tasks to be carried out in discharging the subcontract responsibilities determine the organization that is necessary. The offsite office back up must be tailored to support the on-site tasks. Let us look at influences upon a project's organizational tasks.

It is generally true that building form follows its function, varied by geographical location, e.g. town/city as against a greenfield out of town site where the latter are further influenced by their ability to impact the internal environment via a building structure and fabric designed to integrate with the environmental services. This can mitigate – obviate – the need for mechanical ventilation/air conditioning. Other examples of building differences are:

- pharmaceutical manufacturing requiring segregation of highly serviced sterile/clean areas;
- hospitals with wards 'streets' and operating/treatment areas at the cruciform of 'nucleus' construction;
- shopping centres, highly serviced public areas, low grade services in associated car parks and services connections for tenants in the units and cornerstone development;
- prisons with their highly secure, multiple buildings;
- offices now with structured voice, vision and data cabling;
- leisure centres with wet areas of pools and ice rinks and dry function halls, gyms, snooker rooms, etc.

Add to these the further complexity of the requirement of a phased handover and we have increased management challenge.

Take the prison as a detailed example. It will be low rise, probably of concrete and steel structure, brick clad. There will be a number of prison blocks, bare but highly serviced with toilet, wash handbasins and lighting, all vandal proof with fire, general and security alarms linked to a central control room. There will be reception and discharge blocks, medical, administration and recreational facilities. To manage and record progress the building services contractor will need a flat organogram similar to that shown in Fig. 3.2. For a shopping centre using a multi-service building services contractor the on-site management requirement would look like Fig. 3.3. Not forgetting the smaller jobs such as social housing, with individual small value subcontracts, one for heating, one for plumbing and another for electrics. Then the organogram simplifies to Fig. 3.4.

Clearly the effect of building type and its building services content

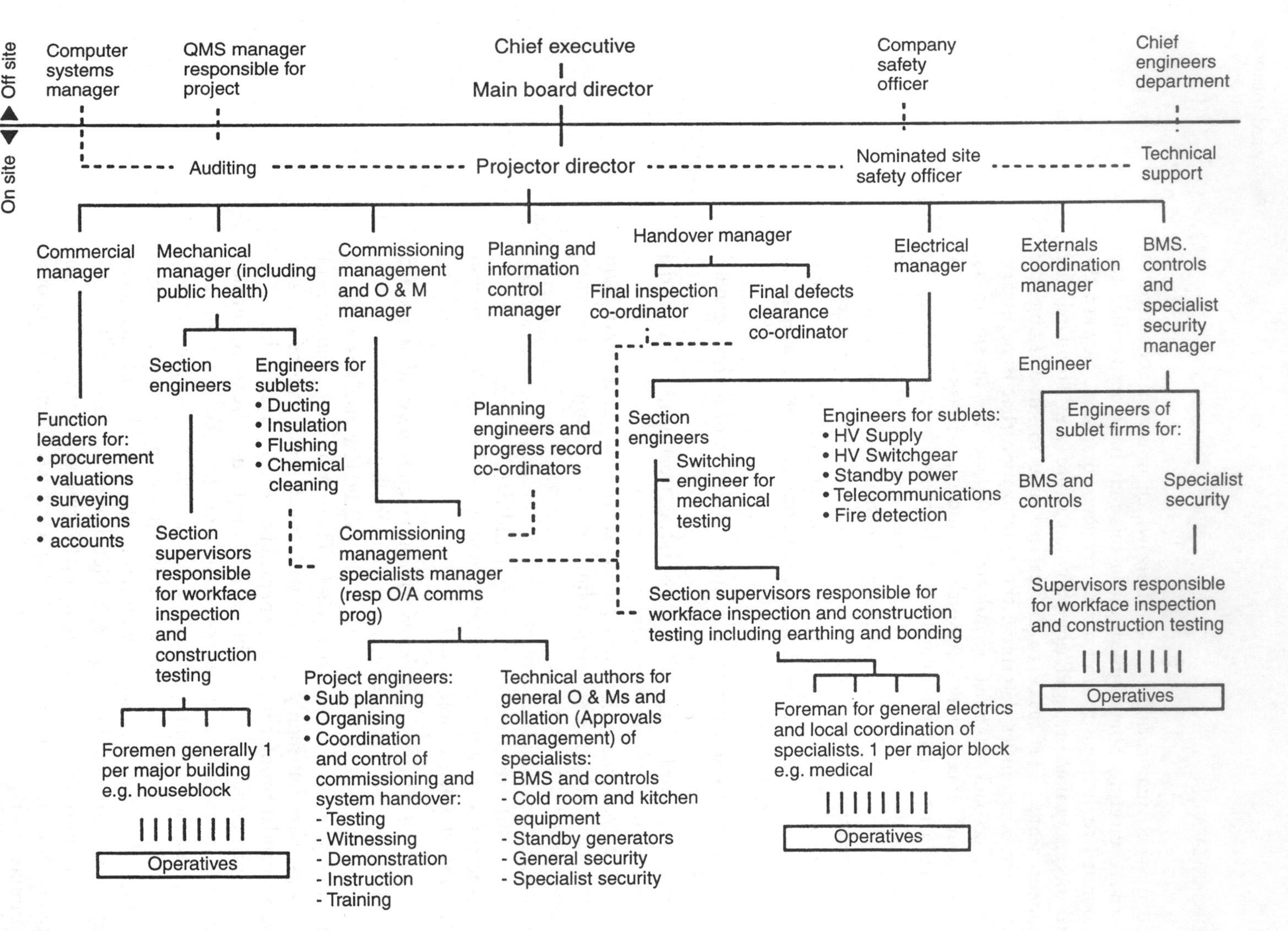

Figure 3.2 Building services contractor – organogram for a new prison.

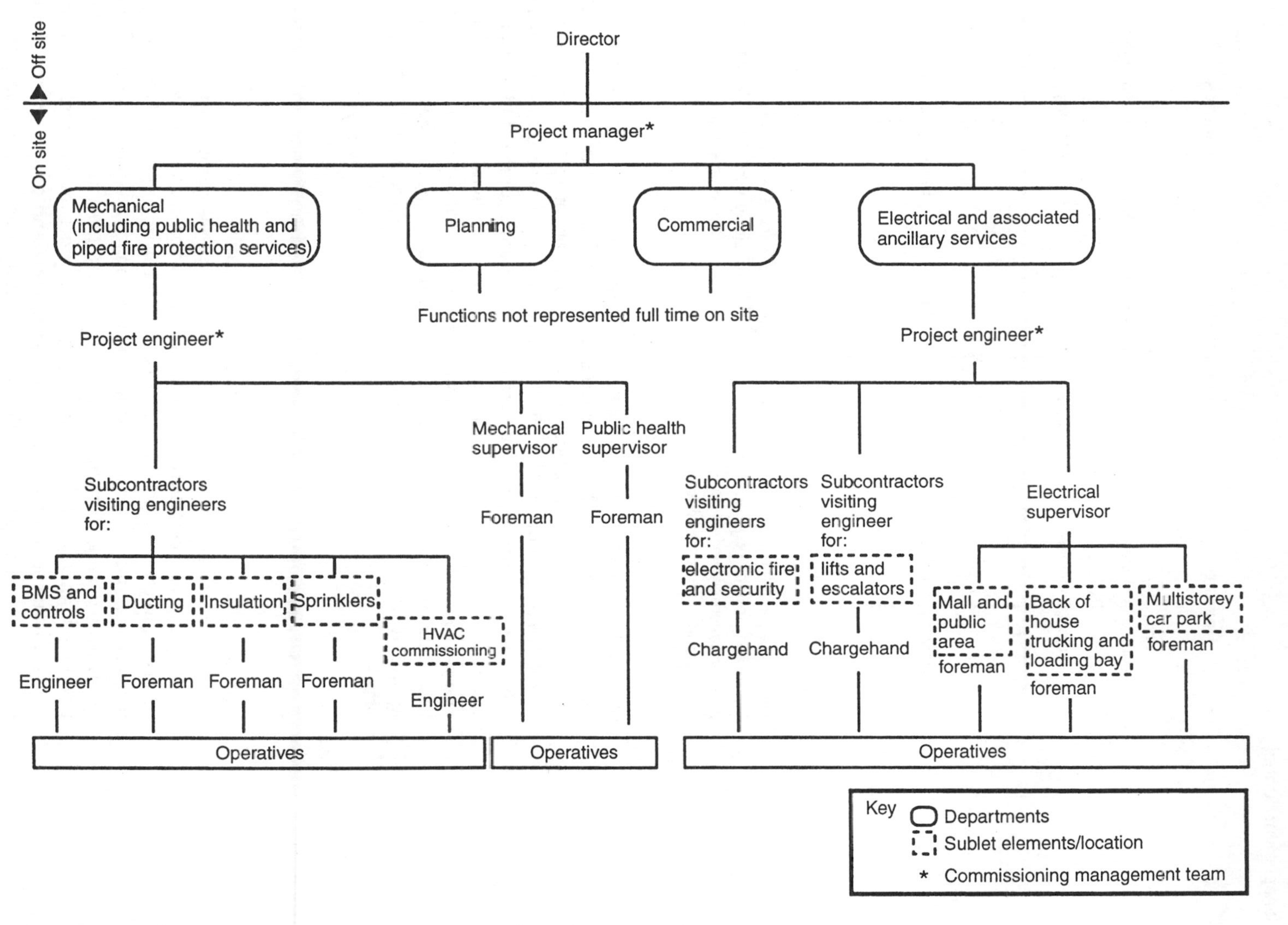

Figure 3.3 Building services contractor – organogram for a shopping centre.

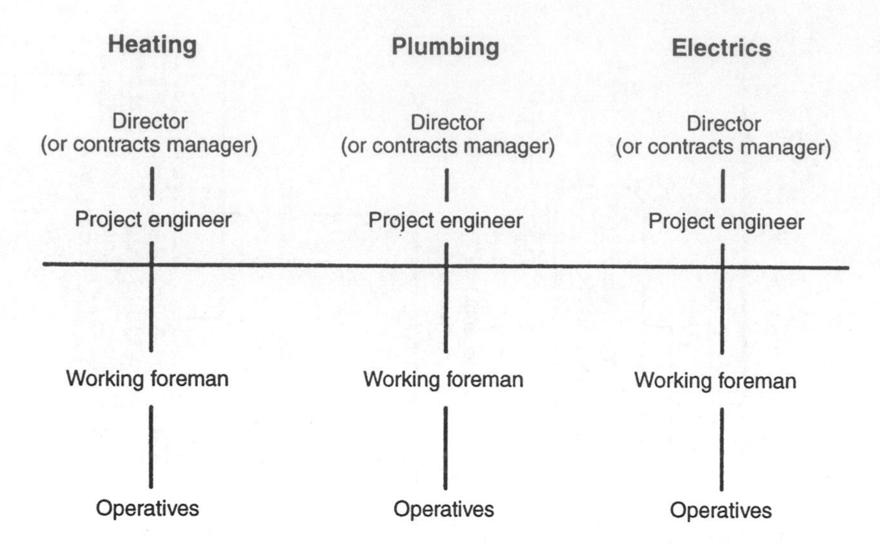

Figure 3.4 Building services contractors' organization for a social housing project. Note: The project engineer may be supported by a visiting supervisor. Whatever line management exists divisions of responsibility must be assigned for inspection, testing and commissioning.

should be reflected in the required management, not all of which is needed until later in the job when specialist subsub contractors on BMS, fire and security, testing and commissioning engineers become critical. These tasks do not involve the heaviest of engineering but are perhaps the most sophisticated. The integration, commissioning and testing of wires, boxes and panels are perhaps the most difficult for a builder to grasp. Paralleled with the need to understand flushing, cleaning, balancing and system proving, earthing and bonding we have jobs that appear to be finished but we are told that that date is still a long way off.

3.3.3 INFORMATION AND DECISION FLOW

Through the on- and offsite management organogram will flow from the client and design team decisions and consequential output in the form of information and instruction to the building constructors. Flowing in the reverse direction will be requests for information, clarification and instruction, and all manner of information in the form of design development, working drawings, programme, samples and mock-ups. The two-way highways will vary in length of physical linkage and use multimedia format – computer disk, fax, hard copy and telephone, etc. All of this the site manager and BS subcontractor must manage and ensure

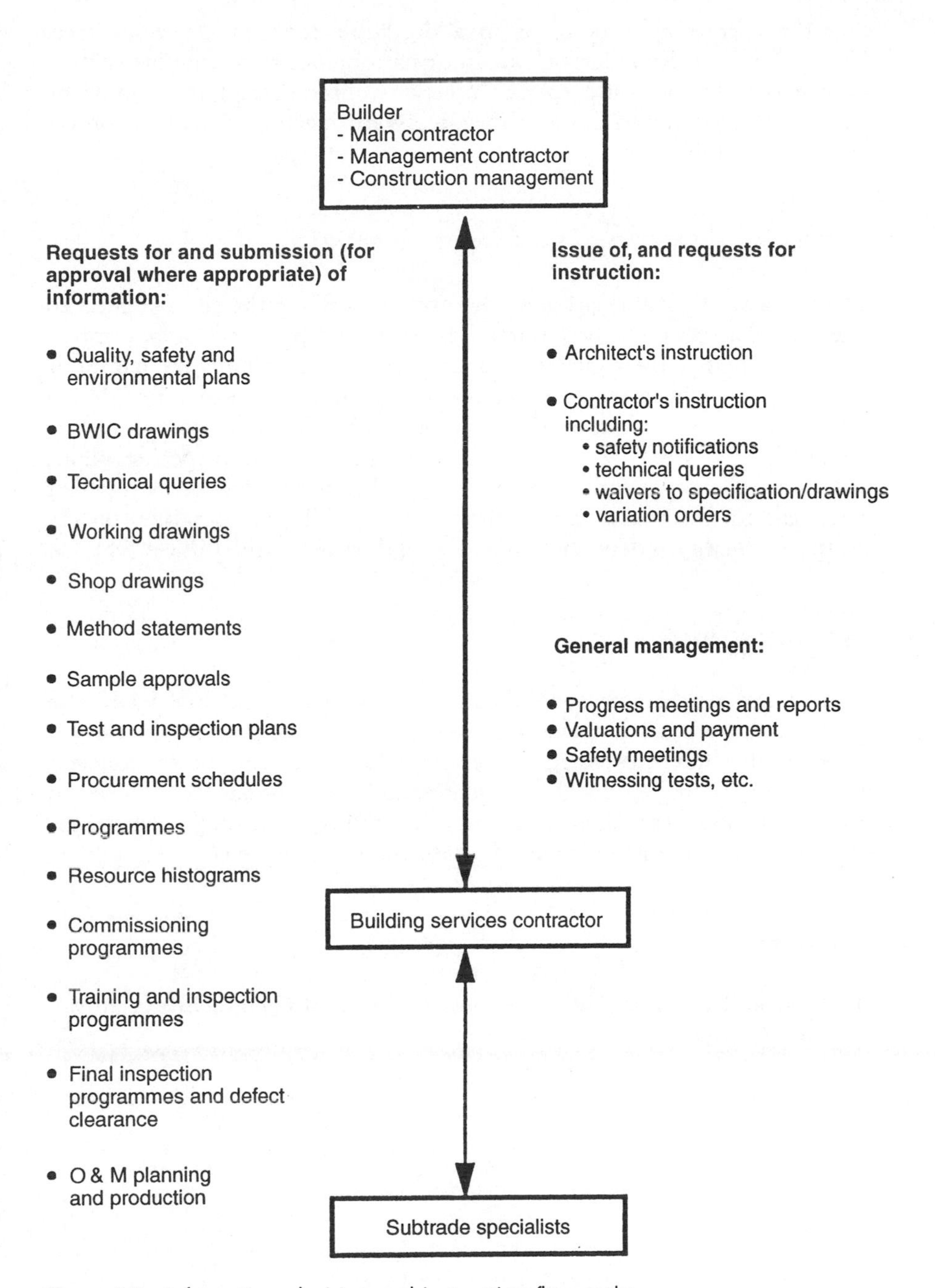

Figure 3.5 Information, decision and instruction flow paths.

that the correct information is in a digestible form at the work face. Figure 3.5 is a reminder of the information, decision and instruction flow paths. The key and specific policy requirements of these must be understood and agreed upon at the pre-award meeting. Details of procedures can follow later.

3.3.4 QUALITY, SAFETY AND ENVIRONMENTAL PLANS

The pre-award meeting presents the opportunity for the site manager to question the services subcontractor on these matters. It is important to find out whether the subcontractor is playing the 'paperwork game' or has a real commitment and understanding of the contractor's, design teams and client's ethos on these matters. The subcontractor should be reminded that what is required are appropriate project specific plans, not churned out standard company material. Ask for the subcontractor's proposals for informing and training his site staff and operatives on the particular quality, safety and environmental issues of the project.

3.3.5 ATTENDANCES

Review the attendances the subcontractor is given and ask what else they will be providing themselves. Asking this question should elicit a response that indicates whether they have looked at the job in sufficient detail. For example, do they understand what fixings they have to provide themselves? This can then be followed through when the subcontractor submits his methods statements for approval.

References

[1] Norman, Donald A. (1992) *Turn Signals are Facial Expressions of Automobiles*, Addison Wesley, Reading, MA.

Quality, safety and environmental plans

4

4.1 General

Up to this point we have been 'constructing the team' and will shortly be joined on site by the BS contractor for the remainder of the contract journey. For the journey to be successful all the way it is essential that the services contractor is properly prepared. The request by the site manager to the services contractor for the preparation of quality, safety and environmental project plans presents an opportunity to make sure that the earlier promises secured at pre-award/management strategy meetings now start to be delivered. It is a very unwise site manager who lightly passes up the opportunity of creating a personal confidence level in the services contractor before that firm has commenced work. Any queries concerning anticipated levels of performance can be dealt with now to avoid facing a retrieval situation soon after site work commences. A shortfall in initial performance by the services contractor is just as much the site manager's responsibility for not expressing doubts when commenting on quality, safety and environmental plans. For clarity the three plans are dealt with separately, although in practice they are more closely entwined.

4.2 Quality plans (QPs)

4.2.1 PURPOSE, PITFALLS AND CONFUSION

Either through market forces or choice most companies have been driven into setting up a quality management system (QMS). Those companies that chose to most fully embrace BS5750 Quality Systems [1] saw the attainment of registered firm status as an integral part of their business plan. Others attained the status but with an ambivalent attitude, waving the certificates wildly until the job was secured, its importance fading during the project. A number of these latter firms received a rude awakening during the surveillance visits of the registering body. For whatever reason and in whatever way and at whatever level quality management systems have been implemented, they are useful, and being useful they are important.

BS4778 Part I 1987, *The Quality Vocabulary Standard* [2], defines a quality plan (QP) as 'a document setting out the specific Quality Practices, Resources, and Sequences of Activities relevant to a particular product, service, contract, or project'. In those 22 words there are a number that are commonly used on a building site. To a site manager they should mean that in calling for a QP he is seeking evidence that the BS contractor will manage the work to a documented system.

The biggest pitfall of QPs is that they are produced, submitted and approved or commented upon, then locked away and ignored. The greatest confusion with project QPs is that they are issued as method statements and vice versa. Look again at the British Standard definition and it is easy to see how the confusion arises for 'setting out the specific ... resources and sequence of activities' is relevant to a method statement. Contractors' familiarity with those words and an insufficient understanding of what is meant by 'a QP' gives rise to the confusion in the minds of both the producer and receiver.

QPs are required to demonstrate that the work to be carried out by the BS contractor will be controlled by the firm's own management system. The site manager, on behalf of his company, needs to be satisfied that the services contractor's QP can be integrated with those of other contractors and that of the project overall.

The site manager should be looking for a QP that gives confidence through its promise of performance. The site manager should use the QP as a tool for monitoring performance and thereby maintaining confidence in the building services contractor.

4.2.2 GETTING APPROPRIATE QPs

For QPs to be useful they must be specific; if they are they will be relevant and appropriate. Figure 4.1 shows the factors influencing QP content. The arrows biasing the content towards greater or lesser detail seem founded in sound logic, but knowledge as to whether the BS are simple or complex has great effect. Look at Fig. 4.2 and take two jobs of the same overall project value, one for social housing, the other a community centre. Both jobs had internal services valued at 20% of the overall project cost. The difference in engineering was extreme. The social housing had domestic-scale plumbing and electrics, with individual low pressure hot water gas fired boilers. The community centre was of squarish deep plan footprint. It comprised a main hall, secondary halls, meeting rooms, creche, kitchen, toilets and staff offices, etc. Many spaces required mechanical ventilation. Heating and hot water was provided from a central gas fired boiler plant. Being a building open to the public it had to comply with far more legislation than the domestic-scale social housing. Fire and security systems were extensive. The main contractor's staffing levels were the same for both jobs and yet the

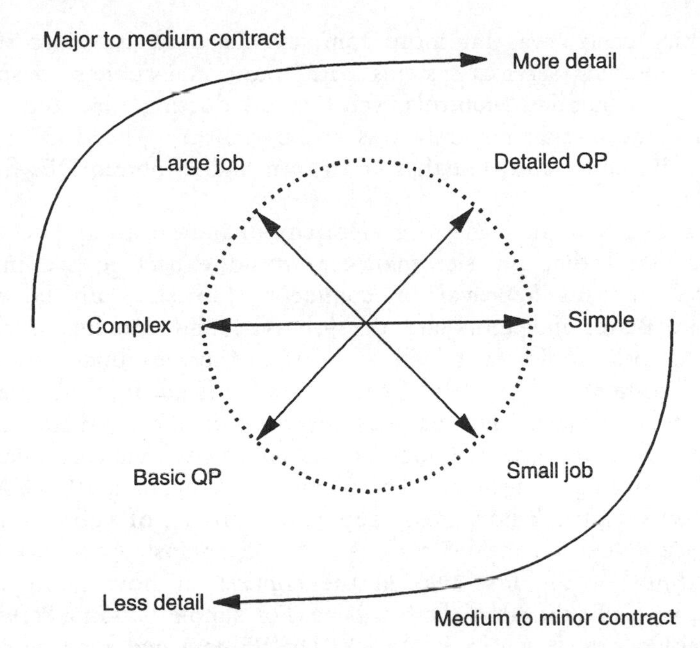

Figure 4.1 Factors influencing quality plan content.

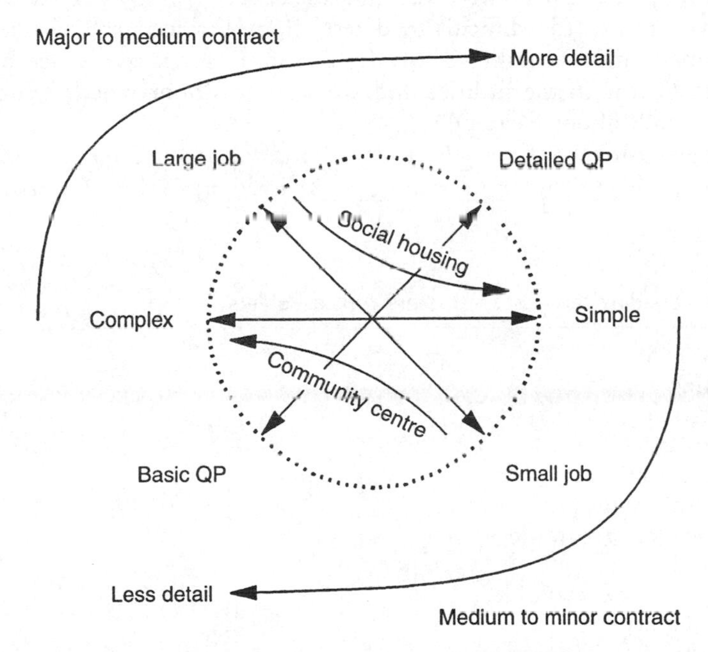

Figure 4.2 Assessing the required level of quality plan detail.
Note: Poor indicator – job size; good indicators – job type, services, content; best indicator – job complexity.

community centre was far more complex. Through its more sophisticated engineering services systems with traditional sublets to specialist traders for insulation, controls, sheet metal ducting, fire and security systems it meant that not only was an all embracing head QP required from the BS contractor, but that he in turn had to obtain QPs from the subtraders.

The above example leads to a hierarchy of indicators in assessing the level of detail that the site manager would expect to see in a QP submitted for his approval or comment. Job size can be a poor indicator. Better indicators are the job type and content of services, while the best indicator of all is born out of an understanding of technical content and number of subtraders involved. It is the number of sublets and specialists that indicates the extent of the organizational and work face interdependencies that the BS contractor has to manage. For an understanding of technical complexity look again at the CAWS in Appendix A and tick how many apply to a project of your own recent experience. Next reconsider the risk profile against Appendix D and finally think of the first two in the context of how many services contractors and subtraders there will be. For simple building services the usual sublets are shown in Table 4.1. The BS contracts may be separate mechanical (M), electrical (E), public health, including sanitation and water (P), sprinklers (F-Fire), and lifts/escalators (L). QPs from each of the above are not too difficult to digest. Complications set in where, as is not uncommon, combined M, E, P and F services are let as one contract. Throw in the utilities and we start to doubt whether it is still correct to talk about basic QPs.

Complex jobs are those wherever, in addition to the basic sublets of Table 4.1, one or more of the services listed in Table 4.2, specialist

Table 4.1 Building services contractors' typical sublets

Mechanical (M)
- Ducting
- Insulation
- Controls
- Commissioning
  - Flushing, chemical cleaning
- O + Ms, record drawings

Electrical (E)
- Lightning protection
- Fire alarms
- Security

Public health (P) (Sanitation and Water)
- Chlorination

Table 4.2 Specialist sublets

– BMS (M) sometimes (E)
– Substation (E)
– High voltage switch gear (E)
– Security (E)
– Data (E)
– Telecoms (E)
– Fire detection and prevention (E)
– Generators (E)
– UPS (E)
– Kitchen and cold rooms (M)
– Process e.g. medical gases
– Commissioning management (M)

services work elements, form part of the jobs content. The designers may have covered the provision of this work as part of the BS without seeking to influence the contractual route by which it should be carried out. If, as is not uncommon, the designer sees these specialist services areas being carried out directly under the main contract the end result as far as QPs are concerned differs only in a little detail. Whether the services are simple or complex what the site manager needs from each BS contractor, be he general or specialist, is a head QP within which are listed all of the QPs the services contractor intends to obtain from his subtraders.

It is unreasonable and not necessary for all of the sub QPs to be submitted at the same time as the directly contracted BS QP. What is important is that the sublet QPs are obtained and approved before work starts on site. It is proving extremely difficult for contractors to obtain appropriate QPs from the utilities companies. This is particularly so with the gas, water and electricity companies. Now in the private domain, they regrettably retain their public utility attitude. One wonders what will happen when a contractor, subject to a BS5750 surveillance visit from the registering authority, is faced with a corrective action because his own QMS states that a QP or quality statement shall be obtained from each subcontractor. Will the utility company still refuse to provide one, putting the contractor's registration in jeopardy, or will some document be produced that will allow the corrective action to be closed out?

4.2.3 SCOPE AND CONTENT

The project QP is a management tool that should be used for the benefit of both building and services contractors. The site manager should be

Table 4.3 Scope and content of building services QP

Authority statement
Control (of contents and issue of QP)
Auditing
Quality statement for project
Management organization (on and off site)
Allocation of responsibilities
Control of documents (including retention of quality records)
Programming and method statements (including access records and lifting)
Procurement (including purchasing)
Control of manufacture (including offsite testing and production)
Packaging, delivery and storage
Product identification and traceability
Supervision and inspection of installation
Testing (of installations under construction)
Commissioning (and final testing)
Protection of the works
Calibration of instruments on site
Handover
– Training and instructions
– Record drawings and wallcharts
– O & M manuals
– Keys and spares
Schedules of standing instructions, project procedures and work instructions
Schedules of company forms
– General proformas
– Commissioning proformas
– Inspection and test sheets

proactive in determining what its contents should cover. The first opportunity to do this would have been at the pre-award or management strategy meeting (see section 3.3). If, as is good practice, the site manager holds a pre-site start meeting, that would have been the second opportunity, providing it was held sufficiently in advance of work commencement. If not, the last opportunity is in comment and approval of the QP submitted. Unfortunately if the views on the document are divergent the seeds of discord are sown. The 'This is inadequate' and 'this is what we usually provide and no one has complained before' phraseology appears. How much better it would have been if both parties had discussed the requirements and provisions. The proactive site manager will have set out his company's requirements with a list of subjects to be addressed along the lines of Table 4.3. For BS the list most closely aligns with the requirements of BS5750, Part 2. What will be received in response will vary in its alignment depending on the value and complexity of the building services content of the project, the size of the BS contractor and their QMS status, i.e. registered or not. The site manager now has to approve the document. He may be assisted in this task according to his own site organizational set up, by visiting or resident quality management and BS staff. Whatever others do for him in approving the QP he should certainly fully understand the reasons for their commentary.

Two examples of quality plans are provided in Appendix H:

1. N.G. Bailey and Co Ltd, quality control plan for the electrical installation at Kingspool development, York (value approximately £1.25 million at 1993) reproduced with kind permission of N.G. Bailey and Co Ltd. This is a good example of the schedule format popular in the industry.
2. Suggested QP for small unregistered services contractors, prepared in descriptive form.

The industry is peppered with many small BS contractors that come between self-employed 714s and those firms of a size that have embraced QMSs and documented them. For the firms below that level many who are competent are seen to be in need of help so that their skills are not forced out of the industry by any incompatible burden of QMS documentation. The QP in Appendix H is therefore an attempt at supporting the small organization by drafting a QP that they can complete, amend and adopt as their own. This last point is extremely important. It must be made clear to the services contractor that the adoption of the QP does not relieve them of their contractual responsibilities.

For an example of a specialist engineering contractors QP refer to the one in the National Association of Lift Makers document *Principles of Planning and Programming a Lift Installation* [3].

4.2.4 A MANAGEMENT TOOL

The site manager who puts the QP away in a file to gather dust has either approved a document that was inappropriate or was unaware of how valuable a management tool a good QP can be. The services contractor has made a statement, and been examined through the approval process, as to how his works will be managed to achieve the specified requirements. Just consider some of the questions that the site manager can raise either to bring a drifting performance back on line, or to be reassured that the journey to completion is going to be a successful one.

- 'I see you were internally audited here last week. Can I have a copy of the corrective actions?'

- 'Why haven't you got a plumbing supervisor as shown on your organization chart?'
- 'I haven't met your safety adviser yet.'
- 'Why was the ductwork delivered last week only found to be damaged in transit two days after it was offloaded?'
- 'Show me that you're keeping a set of marked up drawings.'
- 'Show me the calibration for the pressure test gauges.'
- 'I hear your kitchen equipment installer starts in two weeks. Can I see his quality plan?'

A QP is the framework established at the beginning of the job in an appropriate degree of detail that should be filled out as the job unfolds through planning and programming, installation, commissioning and handover. The message is not to reinvent the wheel at each step, but to use the vehicle already created.

4.3 Safety plans (SPs)

4.3.1 GENERAL

Brought into force on 31 March 1995 the Construction (Design and Management) Regulations (CDM) [4] were far reaching, bringing a co-ordinated 'cradle to grave' approach to health and safety on all but the smallest construction projects. The regulations should be welcomed by every site manager, for no longer is he wholly responsible for constructing someone else's design safely. The CDM Regulations also provide support at the handover stage and beyond. The support comes in the form of a PS appointed by the client, possibly upon the advice of the lead professional or PM.

How well the project is supported by the PS will depend on his or her experience, training and personal characteristics. As this far reaching piece of legislation shakes down and becomes a well rounded *modus operandi* for all parties, we can expect to see some fuzzy edges and sharp corners at the role interfaces of the PS with the DT and site manager.

Most, but not all, of the Factories Act 1961 and Offices, Shops and Railway Premises Act 1963 has been replaced by the Health and Safety at Work etc. Act 1974 (HSW). The Acts form part of criminal law. Under these Acts and common law, employers must have regard to the health and safety of their employees and conduct their business to ensure, so far as is reasonably practical, that contractors, visitors and members of the public are not exposed to personal health and safety risks. The HSW also requires employees to take reasonable care for their own health and safety and of others who may be affected by their acts (or failure to act) at work. To discharge their responsibilities employers must ensure that site managers are competent, i.e. trained in health and safety matters and are provided with a safe place of work. The anomaly here is that it falls to the site manager to provide the safe place of work for his staff and the contractors employed by his company. We will therefore assume that the principal contractor (PC) with overall responsibility for the site under whatever form of contract the project is being constructed has appointed a site manager who has been properly trained. The site manager who has run a number of jobs will have proven his competence. If undertaking his first site manger's assignment then he has his employer's confidence that he will perform competently.

4.3.2 RISKS, HAZARDS AND COMPETENCY

Site manager, were we right to assume your competence above? You are only as competent as the last day of accident free work on your site. If that sounds an unfair statement because others, by their action or inaction, put your project at risk; remember that the accident inspector starts with the person in charge of the site and is reluctant to move down the line of command until he is sure the person in charge is clear of responsibility.

At the beginning of a contract, better still at the beginning of each day, remind yourself of the definitions of risks, hazards and competency. Without a clear understanding of these you are putting yourself at risk. Worse still, you may put others at avoidable risk.

- Hazard: A hazard is anything that can cause harm, e.g. chemicals, electricity, working from ladders, etc.
- Risk is the chance (big or small) of harm actually being done.
- Competent person: In the context of the law requiring some inspection or examination of work to be carried out by a specially appointed person, a competent person is someone who has the necessary technical expertise, training and experience to carry out the examination or test. This could be an outside organization such as an insurance company or other inspecting organization, a self-employed person or one of your own staff who is capable of doing the task. (NB – this is not a legal definition.)

These three definitions were taken from the Health and Safety Executive's (HSE) *Essentials of Health and Safety at Work* [5]. Aimed more at the smaller business the clarity with which it deals with the subject must commend it to every business. Within its 60 pages (the contents list is included in Table 4.4) it creates a framework that can be infilled by the specialist knowledge available from the construction industry, the HSE and British Safety Council. At such a modest cost for its outstanding guidance *The Essentials of Safety at Work* should be on the desk of every project DT Leader, PS and site manager.

Table 4.4 Contents of *The Essentials of Health and Safety at Work* (Source: HSE (HMSO).)

Organizing for safety	
1. Managing health and safety	Some basic information, hazard and risk, where to get help, inspectors and the law
Premises	
2. The workplace	General safety, lighting, hygiene, welfare, display screen equipment, fire precautions
3. Building work	Ladders, scaffolds, roof work, ground work
Plant and machinery	
4. Machinery safety	Guards, machine controls, safe operation, maintenance
5. Gas- and oil-fired equipment	Gas safety, appliances, unburnt fuel and flammable vapours from plant
6. Plant and equipment maintenance	Safe working areas, safe plant, confined spaces, hand tools, vehicle repair
7. Pressurized plant and systems	Design and maintenance, safe operation, pressure cleaning
8. Handling and transportation	Manual handling, repetitive handling, lifting/stacking, transporting
9. Noise	Measuring noise, hearing protection, noise reduction
10. Vibration	Reducing vibration
11. Radiations	Microwaves and radio frequencies, ionising radiations, infrared, ultraviolet, lasers
12. Electricity	Safe installation, insulation, protection and earthing, safe operation, overhead electric lines, maintenance, electric shock, underground cables
Substances	
13. Harmful substances	Control measures, exposure limits, asbestos, lead, skin problems, bacteria and viruses
14. Flammable and explosive substances	Storage, flammable liquids and solids, gas cylinders, dust explosions, oxygen, advice to suppliers, transporting materials
Procedures	
15. Safe systems	Clear and safe procedures, permits to work, lock-off procedures
16. Accidents and emergencies	Emergency procedures, first aid, reporting injuries, investigating events
People	
17. Health care	Health surveillance, mental health and stress, drugs and alcohol, passive smoking
18. Personal protective equipment	Selection and use, maintenance, methods of protection
19. Selection and training	

4.3.3 RISK ASSESSMENT AND MANAGEMENT

Comprehension of the definitions of risk and hazard presents no problem; assessing them is different and requires further help. This help is drawn from Laurence Waterman's article 'What is a "Risk Assessment" Anyway?' [6], in which it is stated:

There are three processes which health and safety law requires:

- Hazard analysis – identifying the hazards in the workplace
- Risk assessment – evaluating the degree of risk, in order to judge what it is reasonably practicable to do
- Risk management – the process by which risks to health and safety are effectively controlled.

Waterman goes on to define three levels of risk:

- The risk is so great or the outcome so serious that it cannot be justified, and therefore the working conditions giving rise to that risk are unacceptable.
- The risk is so small, or can be made so small, that as long as this is maintained, no further action is required.
- The risk lies between these levels, and controls are required to reduce the risk to as low a level as is deemed reasonably practicable. This is where most attention is focused, demanding the intervention of the safety manager, the occupational hygienist, the designer, ergonomist, trainer and a host more. The work represents a risk to health or safety, but the risk is neither negligible nor accepted within the bounds that it be reduced 'so far as is reasonable practicable'.

For the management of risk Waterman advises that the starting point is the following sequence of question and response.

'What are the hazards in the workplace?' Draw up an inventory of challenges to health and risks of accidents. This should encompass not only normal working activities, but also the unusual, contingency arrangements such as breakdowns, maintenance activities, etc. Draw up a checklist for the hazards, using 'brainstorming' and also walk-through surveys. These checklists may then be used periodically by local line managers to check if matters have altered, and new hazards have arisen.

'Are there procedures in place which ensure that the hazards identified do not create unmanaged, uncontrolled risks? Can any risks be eliminated?' If the answer is 'no', then the immediate task of managers is to ensure that the procedures are followed, that staff are informed of the risks and trained to be competent in following the safety instructions. If the answer is 'Yes', then a risk assessment is required – to evaluate what controls are needed.

- Avoid the risks – don't use the dangerous machine or the hazardous substance.

- Combat the risks at source – provide a non-slip flooring in the kitchen.
- Adapt work to the workers – don't expect staff to get used to stretching to reach the controls.
- Use technology – from sound insulation to ventilation.
- Have a coherent policy – don't make *ad hoc* judgements – be organized.
- Protect the workplace – don't rely at all times on personal protective equipment.
- Inform staff – so that everyone understands what s/he must do and why.
- Lead from the top – develop a safety culture from directors to cleaners.

4.3.4 THE CDM REGULATIONS

For a full understanding of the regulations and the responsibilities they place upon the parties involved in a project the Approved Code of Practice (ACoP) *Managing Construction for Health and Safety* [7] would seem to be mandatory reading for the site manager. This ACoP contains good guidance and any failure in compliance might be used as evidence in the event of prosecution. Its sister document *Designing for Health and Safety in Construction* [8] provides advice for designers but the site manager should be aware of its content.

The regulations apply to all but the smallest jobs. By the time the site manager is appointed others in the project team will have decided whether or not the job is of sufficient size for the Regulations to apply. In simple terms for the context of this book the regulations place legal responsibilities on the parties in the following way:

1. The client (including client' agents and developers) must:
   (a) appoint a PS at the concept and feasibility stage;
   (b) appoint a PC.
   The client must ensure the PS, DT and PC are competent and sufficient resources, including time, are allowed for the project to be carried out in compliance with H & S law. For the one-off lonely client discharging this obligation is onerous.
2. The PS must:
   (a) notify the HSE of the project at the concept and feasibility stage;
   (b) report to the client in the concept and feasibility stage on all matters concerning the provision for H & S to be considered during design and construction phases;
   (c) prepare an initial H & S plan during design and planning stage;
   (d) commence the preparation of the H & S file during the design and planning stage;

   (e) confirm to the client that the allocation of resources for H & S are satisfactory;
   (f) at handover pass to the client the project's H & S file.
3. Designers must design in a way which avoids, reduces, or controls risks to H & S as far as is reasonably practicable so that projects they design can be constructed and maintained safely. Where risks remain they have to be stated to the extent necessary to enable reliable performance by a competent contractor.
4. The PC shall:
   (a) manage the project and supervise the activities of other contractors. The PC must ensure that the activities of all contractors are coordinated and that they cooperate each with the other and share relevant H & S information;
   (b) take over, develop and operate the H & S plan;
   (c) ensure contractors they employ are competent and have made adequate provision for H & S requirements;
   (d) set up a procedure so that the views of the workforce can be made known;
   (e) implement site safety training for their employees including the self-employed;
   (f) monitor H & S performance;
   (g) provide the PS with information for the project's H & S file.
5. Contractors shall:
   (a) inform the PC of the hazards of their work and associated risks with assessments and proposals for their control;
   (b) comply with the directions of the PC for the discharge of their responsibilities under the CDM regulations;
   (c) via the PC provide information for inclusion in the project's H & S file.

The object of the CDM regulations is to ensure the safety of a project at all stages in its life as a new, amended, extended, repaired or refurbished construction, including its maintenance. The H & S duties the regulations place on clients, designers and contractors are markedly different from their roles before CDM. A new role is created in the PS, and a PC is defined. They carry the responsibility for the planning, organization, coordination and control of H & S in construction from conception to handover, and beyond with respect to maintenance.

While there can only be one PC his role varies according to the type of contract. Under CM and management contracting the trade and works contractors will operate in the same way as the PC. The CM/MC contractor as PC must have arrangements requesting the trade and works contractors to ensure they appoint subcontractors who are competent. On JCT lump sum it is down to the main contractor as the PC to ensure that his subcontractors *are* competent.

Table 4.5 Contents of the health and safety file and the health and safety plan from a BS viewpoint

---

The health and safety file being primarily concerned with the safe management of constructing or modifying and maintaining a structure should comprise:

- Relevant information provided by the client: for existing buildings this could mean the provision of operating and maintenance manuals and record drawings
  Also to include information on the project site provided by others, such as results of surveys, reports and other investigations
- Details of construction methods and materials used, deemed to include BS and specialist contractors' method statements, risk assessments and safe working practice procedures
- Details of materials and substances used in the construction. To maintain, repair, replace or modify the structure it may be necessary to remove the materials of the fabric, finishes and building services
- Details of plant and equipment installed in the structure. This applies to both BS and Process. Heavy plant and equipment may need to be moved through and in and out of the building for repair, replacement or modification. Movement may be restricted to predetermined routes
- O & M manuals and record drawings.

The health and safety requirements have two stages in their development. The input by the planning supervisor shall comprise the creation of a pre-tender H & S plan and the opening of a H & S file:

- a general description of the construction work
- overall project time with any intermediate stages
- health and safety risks in the construction of the works as identified in the health and safety file
- details and assessments of the competence of designers and contractors
- information necessary to enable any contractor to comply with his H & S provisions. This would include identification of all risks of construction the designers were unable to design out or reduce.

The principal contractor would take the planning supervisor's input and expand the H & S plan to include project arrangements covering the following:

- management and organization to enable all contractors to comply with the applicable regulations of the Management of Health and Safety at Work Regulations 1992 and other relevant statutory provisions for construction work
- ensure compliance with any rules set out in the health and safety plan
- access for persons on site in connection with the project
- display of mandatory notices
- providing the planning supervisor with information that could reasonably be expected to be included in the H & S file rules, in writing
- proposals for informing and training other contractors on the risks to health or safety of that contractor and ensuring they have procedures for passing on that information and training their employees.
- health and safety meetings.

Note: In Clause 2 of the CDM Regulations entitled 'interpretation', 'construction work' is defined in subclause (1): 'The Installation, Commissioning, Maintenance, Repair or Removal of Mechanical, Electrical, Gas, Compressed Air, Hydraulic, Telecommunications, Computer or similar services which are normally fixed within or to a structure' and goes on to define 'Structure' as 'any fixed plant in respect of work which is installation, commissioning, de-commissioning or dismantling and where any such work involves a risk of a person falling more than two metres'. Building services clearly come within these definitions.

In operation the contractual arrangement should have little effect upon what is included in the H & S file. All H & S plans and method statements should be offered up the contractual route for the PS decision on inclusion in the file.

The CDM regulations bring two major documentation changes to the H & S records of a project. The document changes are the creation of the H & S plan, and the H & S file. Both documents are initiated by the PS and remain in that person's control until handed over to the client on project completion.

Table 4.5 summarizes from a BS viewpoint the typical range of contents for (1) the H & S file and (2) the H & S plan.

All parties in the project team comprising, client, designers, PS, PC and contractors will have an input to these documents. The enlightened site manager groaning at the establishment of another channel of communication with a thirst for consuming paper, will feel compensated by the sharing of responsibility. No longer will he have a contracted responsibility for constructing designs that are unsafe in their mutation under construction, from temporary to permanent works. The site manager is now in a more powerful position to obtain safety plans and worthwhile method statements from contractors, which must include the utilities companies.

4.3.5 LEGISLATION IN CONNECTION WITH BUILDING SERVICES

From the 'catch all' of the CDM legislation we will look at the Acts and enabling Regulations that affect the designer and the installing contractor. The design will be dealt with in general terms and construction in some detail.

The services designer is responsible for the building services complying with the current Building Regulations, Fire Regulations and the Regulations of the HSW. In addition there will be requirements of the client's insurance company to be complied with. To guide the designer the CIBSE has published a Technical Memorandum (TM 20) *Health, Safety and Welfare – Guidance for Building Services Engineers* [9]. Whether they are engaged in the design, construction, installation, commissioning, operation or maintenance of building services systems the guide makes them aware of their professional responsibilities under the HSW and other relevant legislation. 'This publication offers guidance on the legislation which applies in particular circumstances, illustrates the risk to which persons at work may be exposed and identifies authoritative sources of information to enable adequate standards of Health and Safety to be maintained.' It is an important reference as it brings together all stages in a project's lifecycle. For our simple purpose design responsibility is separated from construction. For D & B contractors this will never be the case.

Many PCs have, for a number of years, implemented the preparation of a project safety plan that is put in place before work commences on site. BS contractors trading as PCs, particularly on refurbishment projects (due to the high value – up to 70% or more, of building services), will find it essential to produce a project safety plan.

Table 4.6 has been compiled from the Heating and Ventilating Joint Safety Committee 1994 Document, *Site Safety: A Guide to Legal Responsibilities for Health, Safety and Welfare* [10]; it is another

Table 4.6 Health and Safety legislation related to the installation of building services

1. Health and Safety at Work Act etc. 1974
2. Management of Health, Safety and Welfare Regulations 1992
3. Construction Regulations 1961 and 1966
4. Control of Substances Hazardous to Health Regulations 1994
5. The Reporting of Injuries, Diseases and Dangerous Occurrences Regulations 1985
6. Work Equipment Regulations 1992
7. Manual Handling Regulations 1992
8. Personal Protective Equipment Regulations 1992
9. VDU Regulations 1992
10. Noise at Work Regulations 1989
11. Electricity at Work Regulations 1989
12. Health and Safety (First Aid) Regulations 1981
13. Ionising Radiations Regulations 1985
14. Safety Signs Regulations (Amended) 1980
15. Gas Safety (Installation and Use) Regulations 1994
16. Construction (Design and Management) Regulations 1994
17. Environment Protection Act 1990
18. Employer's Liability (Defective Equipment) Act 1969
19. Employer's Liability (Compulsory Insurance) Act 1969
20. Trade Union Reform and Employment Rights Act 1993
21. Abrasive Wheels Regulations 1970
22. Asbestos (Licensing) Regulations 1983
23. Control of Asbestos at Work Regulations 1987 and Amendment Regulations 1992
24. Control of Lead Regulations 1981
25. Highly Flammable Liquids and Liquefied Petroleum Gas Regulations 1972
26. Safety Representatives and Safety Committee Regulations 1977
27. Road Traffic (Carriage of Dangerous Substances in Packages, etc.) Regulations 1986
28. Fire Certificates (Special Premises) Regulations 1976
29. Control of Pesticides Regulations 1986

Note: The list is not intended to be exhaustive but indicative of the legislation applicable on most BS projects. Work in the agricultural, pharmaceutical, health care, food processing and any of the special industries would add further legislation to the list.

excellent source of reference from the BS industry. The site manager's own company may have procedures in their H & S manual for all of this legislation. It will be noted how much common ground there is between the builder and the services contractor. Where the services contractor is working under a CM or MC contractual arrangement and providing much of his own attendances and builders work, the similarity in safety responsibilities is even closer.

4.3.6 SCOPE AND CONTENT OF BS SAFETY PLANS

The CDM regulations have established a communications highway with respect to H & S information. The PC will have received information to be taken into account when tendering and producing his project safety

Table 4.7 Suggested project safety plan contents list for a building services subcontract

1.	Introduction	Description of project
2.	Objectives	The implementation of appropriate safety procedures to avoid accidents, dangerous occurrences and create a safe place of work
3.	Control documents	Refers to the specific document that the contractor will be associating with the safety plan e.g. company safety instructions, safe working practice method statements, company procedures for COSHH and perhaps HSE guidance notes, HVCA, ECA and NALM documents
4.	Allocation of responsibilities	An organogram supported by a description of heirarchal function and responsibility for safety management
5.	Subcontractors specialists	A list of trade and specialist work to be subcontracted which will be amended by the addition of names as appointments are made. A description of the building services contractors management, i.e. what he will be asking them to provide e.g. safety plans similar to his own but appropriate to their work
6.	Work instructions	List to be used (if they already exist) or to be drawn up specific to the project
7.	Safety instructions	A description of the contractor's proposals of safety inspection and auditing
8.	Storage of materials	particularly describing safety precautions
9.	Control of waste and rubbish	This may be correlated or covered in a separate environmental plan if one has to be provided
10.	Records	Details of folders and location of statutory registers, accident book and health and safety checks
11.	Safety equipment	Describing personal protective equipment (PPE), its use and control
12.	Safety training	Describing proposals for the project, how and by whom, when and what records will be held
13.	Review and amendment	Programme of review, amendment and reissue of the safety plan
14.	Appendices	Lists of documents forming part of the safety plan, e.g. organogram, details of subcontractors, lists of work instructions, audit programmes, etc., etc.

plan. Similarly it is essential for the communication route to be extended to all contractors. It is not unreasonable for the BS contractor to receive information and respond to it in the same way, and so on down the line to subtraders and specialists. There are strong analogies here to the provision of QPs, one being 'buttoned in' to the one above from the bottom (least significant contractor) to the top, the client. Table 4.7 suggests a BS SP contents list.

Safety plans, like QPs, are dynamic documents subject to amendment. Changes in practice arising from reviews and audits will create amendments, as will the introduction on to site of every sublet or specialist having been enabled by an appropriate QP, safety plan and work method statement. As for QPs, the chain of interlinked SPs will be reflected in its complexity by the subcontract grouping of M, E, P, F and L and the sublets involved. Where the building services work is carried out by a multi-services contractor some of the onus passes from the site manager to that leading services contractor. In the role of employer the multi-services contractor assumes many of the responsibilities of the PC. As it was with QPs so it is with safety plans. The multi-services contractor will identify the particular H & S risk for the general BS work, pointing up that initial information which will be updated as each subtrader comes on board. For the multi-services contractor what we then have is a general safety plan which by the end of a project has been expanded into a detailed plan. A general safety plan will refer to general method statements where the requirements for further risk assessment proposals, precautions and monitoring procedures should be found. Having noted the areas of legislation they have in common (see Table 4.6) we will concentrate on the differences between the services and building work elements.

4.3.7 GUIDANCE TO UNDERSTANDING AND COMMENTING ON BS SAFETY PLANS

For the large range of activities that are common to both building and services work check that the services contractor's safety arrangement procedures are as good as your own company's. If the services contractor's information is inadequate recommend that they adopt your company's, but advise them that they do so without changing their contractual and legal responsibilities.

If the safety plan has been submitted with method statements, ask the services subcontractor to confirm what risks have been identified and assessed. The site manager should be responsible for carrying out his own risk assessments. This can only be done with job knowledge. In doing this consider the following:

- The relationship of an activity to its location defines the risk.

Table 4.8 Key risks and controls required for building services activities at different stages of the construction process

Stage of process	Activity	Key risk	Controls
Construction	Positioning plant and equipment	Lifting, hoisting, handling (including moving through a building)	Safe system of working
	Erecting plant and equipment	Welding chemicals	
	Building pipe, ducting and electrical distribution systems	Welding chemicals, pressure, temporary works	Permit to work Hot work permit
Preparation of systems	Flushing Chemical cleaning Water treatment Blowing out ducting	Temporary works, pressure Chemicals Dust	
Commissioning	Regulation of systems and setting to work	Live plant, equipment and systems	Permit to work Notification to others (systems working under test) Access permits
Operation	Operation of systems for: Arranging tests e.g. fire officers/ insurers Witnessing specified function/modes Training and instruction	Live plant, equipment and systems	Notification to others Appointment of competent person(s) Access permits Operation manuals available

Note: All activities should be carried out to approved methods in which hazard and risk have been identified and a safe system of working agreed upon. The Key Risks and Controls are those for which Building Services are recognized as being different from the main stream of other building activities.

- All risks should be managed through proper supervision of a trained workforce.
- Each activity of constructing a building services system has its own range of risk (see Table 4.8).

The major building services contracting associations give guidance to their members on risk assessment. The HVCA brought in consultants to produce its *Risk Assessment Manual* [11]. The Electrical Contractors Association (ECA) found Hascom Network Ltd with a ready made system which over a few weeks was tailored to the requirements of electrical contracting. The ECA recommends its members use the *Hascom Manual* [12]. The objective of both associations is to help their membership comply with the requirements of the Management of Health and Safety of Work etc. Regulations 1992. The Hascom Manual proved so successful that the ECA were requested to provide more risk assessments so that it could be used by the construction industry generally. The Hascom/ECA approach has been to produce a generic risk assessment form on the reverse of which is the site-specific assessment. Table 4.9 lists generic risk assessments for electrical contracts.

Table 4.9 Generic risk assessments for electrical contractors (ECA *Risk Assessment Manual and Supplement*) (Copyright 1996 Hascom Network Limited.)

EC1	Working at height
EC2	Use of access scaffolding
EC3	Use of mobile scaffold towers
EC4	Use of ladders
EC5	Use of trestles
EC6	Use of mobile elevating working platforms
EC7	Storage of materials on site
EC8	Storage and use of LPG
EC9	Storage and use of highly flammable liquids
EC10	Slinging of loads
EC11	Use of lifting equipment
EC12	Lifting operations using mobile crane
EC13	Use of materials hoist
EC14	Use of disc cutters and abrasive wheels
EC15	Use of cartridge operated fixing tools
EC16	Use of portable pipe threading machines
EC17	Use of hand tools
EC18	Use of compressors and pneumatic power tools
EC19	Use of vertical drilling machines
EC20	Use of fork lift trucks
EC21	Use of portable electrical equipment
EC22	Installation and use of temporary electrical supplies
EC23	Installation of cable trunking and trays
EC24	Chasing out for cable runs
EC25	Electrical work – up to 415 volts
EC26	Installing/replacing luminaires
EC27	Cable pulling
EC28	Electrical testing and commissioning
EC29	Charging/servicing electrical batteries
EC30	Work on equipment containing PCBs
EC31	Disposal of fluorescent luminaires
EC32	Disposal of waste materials

Table 4.9 *Continued*

EC33	Work in confined places
EC34	Work near or under overhead power lines
EC35	Work in the vicinity of underground services
EC36	Work with lead and lead compounds
EC37	Work involving asbestos-containing materials
EC38	Working alone
EC39	Work in occupied premises
EC40	Work on fragile roofs
EC41	Work in and with excavations
EC42	Use of excavators
EC43	Use of small dumpers
EC44	Excavators used for lifting
EC45	Setting up site facilities: offices, welfare and storage
EC46	Fire on site
EC47	Work in electrical workshops
EC48	Office work
EC49	Driving company vehicles
EC50	Erection and use of falsework
EC51	Work on or near water
EC52	Work on live sewage connections in shallow excavations
EC53	Pipe soldering
EC54	Pressure testing
EC55	Pipework installation
EC56	Flue liner installation
EC57	Fan coil installation
EC58	Plumbing
EC59	Use of arc welding equipment
EC60	Use of gas welding/cutting equipment
EC61	Dismantling/installing ductwork
EC62	Cleaning grease extractors
EC63	Work with flat glass on site
EC64	Use of skips
EC65	Work with non-asbestos insulation materials
EC66	Work in joinery workshops
EC67	Roadworks
EC68	Road transport on site
EC69	Operating road surfacing plant
EC70	Minor demolition and breaking out of surfaces
EC71	Work in cofferdams
EC72	Use of high-pressure water/steam cleaners
EC73	Use of bench-mounted grinding machines
EC74	Use of portable woodworking machines
EC75	See also EC65
EC76	Use of bitumen boilers
EC77	Use of lasers (Class 1 and 2)
EC78	Use of earthmoving plant

GENERIC RISK ASSESSMENT

ACTIVITY COVERED BY THIS ASSESSMENT	ELECTRICAL WORK - UP TO 415 VOLTS			
SIGNIFICANT HAZARDS	**ASSESSMENT OF RISK**			
	INSIGNIFICANT	LOW	MED	HIGH
1. Electrocution				X
2. Electrical burns				X
3. Fire			X	
4.				
5.				
6.				
7.				
8.				
9.				
10				

ACTIONS ALREADY TAKEN TO REDUCE THE RISKS:
Compliance with:

Provision and Use of Work Equipment Regulations 1992
Management of Health and Safety at Work Regulations 1992
Electricity at Work Regulations 1989 (EAW)
IEE Wiring Regulations, 16th Edition and guidance
HSE Booklets: HS(R)25 - Memorandum of guidance, EAW Regulations
HS(G)85 - EAW - safe working practices
HS(G)13 - Electrical Testing (out of print)
HSE Guidance Notes: GS38 - Test equipment for use by electricians

BS 6423:1983 - Code of practice for maintenance of switchgear & control gear up to 1kV.

Planning:
Whenever possible, 'live'working is to be avoided. If 'live' work is required the assessment procedure in Figure 1 of HS(G)85 is to be followed and a safe system of work devised, preferably in writing. Sufficient personal protective equipment is to be available in the workplace.

Physical:
Access to live conductors is to be controlled, and appropriate signs are to be in place. Written information and instructions will be required for work on complex systems (control, metering & parallel circuits). A clear access of 1m, gloves and matting to BS697 and BS921 are to be provided for 'live' working. Electrical test equipment will be insulated and fused to GS38 requirements and in date for calibration. Electricity supply authority seals will not be broken, and final connections will not be made without written authority.

All circuits to be worked on will be treated as live until verified dead. There are no exceptions to this requirement; experience of employees is irrelevant.

Managerial/Supervisory:
Live work is only to be carried out by authorised competent electricians under direct supervision of nominated supervisors. Electricians will not be permitted to work unaccompanied on live connections above 125 volts unless specifically authorised to do so, and good communications are in place. Adequate PPE, first-aid and qualified first-aiders are to be available at the workplace where live work is to be done.

Training:
The qualifications and competence of all persons carrying out electrical work will be verified by inspection of current certificates held of training/experience. Before authorisation, operatives will be trained in the IEE Wiring Regulations 16th Edition, and the Electricity at Work Regulations and guidance. Before authorisation to carry out 'live' work, they will be trained in the safe working practices contained in HS(G)85, and in any written safe work systems. All electricians will be trained in the treatment of electric shock and burns.

THIS GENERIC ASSESSMENT MUST BE COMPLETED BY ADDITION OF SPECIFIC SITE DETAILS ON THE REVERSE	FILE REFERENCE: EC25

Copyright 1996 HASCOM NETWORK LIMITED ELECTRICAL CONTRACTOR'S ASSOCIATION

Figure 4.3 Generic risk assessment form by Hascom Network Ltd for electrical work up to 415 volts.

SITE-SPECIFIC ASSESSMENT

On each site and each location, the generic assessment overleaf must be reviewed to ensure that all significant hazards and their risks are identified and controlled. Completion of this side will ensure that your assessment is both appropriate and complete.

SITE/LOCATION:

MAXIMUM NUMBER OF PEOPLE INVOLVED IN ACTIVITY:

FREQUENCY AND DURATION OF ACTIVITY:

ADDITIONAL SPECIFIC HAZARDS IDENTIFIED:

ADDITIONAL CONTROL MEASURES REQUIRED:

ASSESSMENT OF REMAINING RISKS: Insignificant / Low / Medium / High

SERIOUS & IMMINENT DANGER RISK IDENTIFIED: Yes / No

EMERGENCY ACTION REQUIRED:

NAME(S) OF COMPETENT PERSON(S) APPOINTED TO TAKE ACTION:

CIRCUMSTANCES WHICH WILL REQUIRE ADDITIONAL ASSESSMENT:

1. Any work on circuits involving voltages in excess of 415 volts
2. Any work on equipment which may be energised by third parties

CIRCULATION OF RISK ASSESSMENT:

CONTRACTOR	☐	SITE COPY	☐	EMPLOYEES	☐
SUBCONTRACTOR	☐	OTHER OCCUPIER OF PREMISES	☐	CLIENT	☐

ON-SITE ASSESSMENT SIGNED:	DATE:	EC25B

Copyright 1996 HASCOM NETWORK LIMITED ELECTRICAL CONTRACTOR'S ASSOCIATION

Figure 4.4 Site-specific assessment form by Hascom Network Ltd for electrical work up to 415 volts.

The assessment sheets for electrical work up to 415 volts are included as Figs. 4.3 and 4.4 by kind permission of Hascom.

The National Association of Plumbers provide information to their members through Health and Safety Sheets No. 1 and 2 which include the well known 'Sharps' procedures.

The BS professional institutions, research organizations and trade associations take safety matters seriously and have created appropriate guidance for their members. Throughout the range of building services risks to be assessed there are a plethora of substances whose use is hazardous to health. For example, in preparing and jointing pipework of all materials the preparation and bonding involves the use of chemicals in the form of pastes, solvents and welding rods. Many are dangerous when applied in free air conditions, becoming lethal when used in confined spaces of ducts, crawlways and corners of plant rooms, which are the 'normal' places of work for building services operatives. All materials used on site can, according to their composition and method of use, be subject to the Control of Substances Hazardous to Health Regulations (COSHH) 1994 [13]. As a piece of legislation COSHH stands alongside the CDM regulations in importance to the construction industry. The HVCA responded to this legislation with the production of its *COSHH Manual* Volumes 1 and 2 [14], running to nearly five hundred pages and covering two hundred materials. Volume 1 contains information on implementing the regulations, obtaining information from suppliers, main and subcontractors and guidance on site supervision.

There is no shortage of published information from authoritative sources that can aid the site manager in assessing a services contractor's safety plan. Of course, it is not intended that the site manager should take responsibility away from the services contractor, but he may wish to spot check a few items, or at least see the light of recognition in the eyes of the contractor when asked if any of the trade association documents mentioned have been referred to. Having catalogued very broadly the range of risks in building services these will be returned to in the contexts of other chapters, particularly Chapter 7 on supervision and inspection, and Chapter 9 on commissioning and its management.

4.3.8 TRAINING

Under Regulations 17 and 18 of CDM it is a requirement of the PC to provide training and guidance to all employees with regard to H & S. This refers specifically to the PC's own employees, but it is the wise contractor who has made it a requirement of his enquiry to the building services contractors that they also shall provide training and education for their employees. The building services contractor is usually required

Table 4.10 Toolbox talks 1–24 (Source: HVJSC.)

1. Safety awareness	13. Noise at work
2. Good housekeeping	14. Use of respirators and breathing masks
3. Safety signs and notices	15. Substances hazardous to health
4. Manual handling of loads	16. Working in confined places
5. Working with hand tools	17. Working with electricity
6. Wearing and caring for PPE	18. Lifting operations
7. Working at heights 1. ladders	19. Lifting gear
8. Working at heights 2. step ladders	20. Permit to work
9. Working at heights 3. trestle scaffolds	21. Working outdoors
10. Working at heights 4. tower scaffolds	22. Fire prevention on sites
11. Working at heights 5. Scaffolding	23. Working in excavations
12. Working on roofs	24. Fire on site

to attend regular site safety meetings with other contractors. The recognition and willingness to attend the PC's meetings and provide on-site training and education should be in the BS safety plan.

The Heating and Ventilating Joint Safety Committee has produced the 24 toolbox talks listed in Table 4.10. These talks are intended to be of 10–15 minutes duration, and presented without specialist knowledge on the part of the speaker. They cover the most basic of H & S points concerning topics, activities and locations. By way of example, toolbox talk No. 20. 'Permit to work systems' is reproduced as Table 4.11 with kind permission of the HVJSC.

NALM provide a distance learning course, Unit 9, Safety and health: managing safely. Aimed at all from directors to fitters and testers a synopsis of the course is shown in Table 4.12. Training along these lines provided through trade associations will meet the requirements of CDM. The site manager might also enquire of the services contractor what personal safety guides are issued to employees; one example of these is the HVJSC's *H & V Safety Guide* [15].

It should boost the site manager's confidence if the services contractor is a member of the appropriate trade association and all individual operatives are accredited by the same organization. Both ECA and HVCA are pushing hard for the accreditation of individuals, an attitude that has the implied approval of the Latham Report [16]. The HVCA register of operatives in the heating, ventilating, air conditioning and refrigeration industry certifies qualifications and training. It should be noted that operatives with welding competency may be HVCA registered while holding the competency certificates of other authorities such as the Heating, Ventilating and Domestic Joint Industrial Council.

Table 4.11 Toolbox talk No. 20 (Source: Heating and Ventilating Joint Safety Committee.)

A Permit to work systems

B The system of issuing Permits to Work was first-introduced in high risk industries like mining and petrochemicals. The improvement in safety was so marked that the practice has been extended to all industries where a task involving a special risk is to be undertaken. Examples of circumstances where a Permit to Work system may beneficially be operated are:

(a) Electrical Work
(b) Roof Work
(c) Trench Work
(d) Hot Work
(e) Confined Space Work
(f) Work near or above deep water
(g) Work in radiation 'controlled areas'.

C No special legislation requires Permits to Work, they are just a way of ensuring a strictly controlled Safe Place of Work and a Safe System of Work in difficult circumstances. They also allow supervisors to keep a check on what is happening by limiting the issue of permits to what can actually be supervised. A Permit to Work will often be accompanied by a Method Statement stating how the job is to be done.

D A Permit to Work ensures that:

1. The task to be done is clearly stated.
2. All potential hazards have been considered and the risks assessed.
3. The measures appropriate to eliminate or control the risks have been put in place.
4. The person(s) to do the work are clear about it, and the safety precautions to be observed.
5. The person authorising them to do it is satisfied about the safety of the task and method of working.
6. The date and time when the work is to be done is agreed and also when work will stop (finished or otherwise).
7. The person authorising the work is told, when the work stops, what stage the job has reached e.g. 100% finished; – 50% finished etc.
8. The person acknowledges that he has been told what state the plant is in e.g. ready to run; – further work needed etc.

What to do upon receipt of a Permit to Work

1. Check that all sections have been completed i.e. all hazards have been considered.
2. Check the date and times when the permit starts and expires. Note: Permits are issued to individuals, therefore should only be valid for one shift. Circumstances can change while you are away, so a new permit is necessary for the next shift.
3. Check the work location to ensure that no problems have been overlooked. Check that persons not included on the Permit are excluded from the area (barriers, notices, etc.).
4. When you are sure everything is in order, sign for acceptance of the permit and commence work.

During the Work

1. Ensure that everybody involved observes all the conditions of the permit. Do not relax any of the stipulated precautions.
2. Make sure any safety devices like padlock keys or fuse links are safely in your possession.
3. If things go wrong or the situation changes notify the authorised person at once. The Permit may need to be cancelled and a new one issued to cover the new situation.

Table 4.11 *Continued*

4. If time runs out, stop work and notify the authorised person at once. He can decide to issue a new Permit or to extend the time.
On Completion
1. Return the Permit to the authorized person and both of you sign it to show that the work is complete and the responsibility is passed back to him.
E Any Questions
F Summarise Main Points

All who work on gas installations, whether firms or individuals, must be registered by the Confederation of Registered Gas Installers (CORGI).

Competent electricians may be able to demonstrate that they are on the voluntary register of the Joint Industry Board's List.

If the site manager is in any doubt about competency of a building services operative he should call for evidence of that person's registration.

4.3.9 SITE MANAGER'S RESPONSIBILITIES FOR SUBCONTRACTORS ON SITE

The CDM regulations have done little to change the responsibilities of a building contractor for his subcontractors. The openness of the communications highway down which the evidence of discharging this responsibility will pass on its way to the PS is doing much to enhance a previous area of weakness. In October 1984 the Building Employers Confederation published *A Guidance Note on Safety Responsibilities for Sub-Contractors on Site* [17]. The author believes that the CDM Regulations override the BEC document in the following two ways:

1. making the contractor's role more onerous by changing from 'assisting subcontractors to work safely' (Clause 1.1) to ensuring they comply with relevant Health and Safety legislation;
2. 'subcontractors should ask to see the main contractor's statement (H & S policy issued to employees) and those of other relevant contractors on site, so that each will know the other's organization and arrangements for the H & S of employees.' (Clause 2.1) It will be difficult for PCs to come to terms with this long dormant recommendation from their own organization. Openness and compatibility

Table 4.12 NALM distance learning, Unit 9: Safety and health – managing safely

A.	General approach	Principles of management as applied to safety and health Safety policies, responsibilities Risk management principles, cost/benefit estimates Quality in safety and health management Role and function of safety and health practitioners
B.	Understanding why incidents occur	Immediate and underlying causes Training, supervision Near misses
C.	Applying management principles to safety and health	Identification of hazards and assessment of risk Strategies for controlling risks Evaluation of performance
D.	Behavioural factors affecting safety and health performance	Physical factors Psychological factors
E.	The legal framework for safety and health at work	General duties under Sections 2–6 of the Health and Safety at Work Act 1974 Responsibilities of employers under the EC 'framework' directive Improvement and prohibition notices System of health and safety regulations Fire precautions legislation and its enforcement Civil liability for compensation for injuries
F.	Communication skills in relation to safety and health	Principles of communication Safety and health reports
G.	Common occurring hazards	Access and egress – general, floors, passages, stairs and lift landings Machinery – general, guarding particularly machine/pulley room Electrical – equipment, installations, 'live' working and permits to work Manual handling – assessments, information, selection and training Mechanical handling – general, lifting equipment, slinging, and operation Health – substances, noise, temperature, vibration and repetitive movements Fire – precautions, procedures including hot work Waste – duty of care and procedures PPE – types, selection and use (including associated hazards)

Table 4.12 *Continued*

H.	Industry specific hazards	Working at heights – equipment including scaffold, ladders and personal fall protection Working alone – identification, procedures (including emergencies) and training Confined spaces – safe working procedures, permits to work and training Tools and equipment – general, hand, cartridge and power tools, welding and burning Traction and hydraulic lifts – general safe working procedures and training Observation lifts – general, safe working procedures and training Partially enclosed lifts – general, safe working procedures and training Escalators – general, safe working procedures and training

between organizations should make steps towards improved harmony thereby reducing risk to personal safety on site. The CDM Regulations are a commendable route to the achievement of this objective.

4.4 Environmental plans (EPs)

4.4.1 GENERAL

A considerable amount, perhaps the larger part, of construction work carried out in the mid-1990s, has been with some QMS in place. Even if not going all the way to BS5750 registration a great number of builders, trade contractors and specialists have taken quality building into their business ethos. If QMS is not mandatory CDM certainly is. Together these two influences have made an important impact upon the industry. Now they are being joined by environmental management systems as serious consideration is given to the effect upon the environment of constructing buildings. As with QMS there is the same push and pull in businesses putting in place environmental management systems.

This is not the place to expand the issues and ride personal hobby horses. It is sufficient to say that from the 1992 Rio de Janeiro International Conference on the Environment, pressures for better care of the environment have increased from European down to local community levels. The European Community does seem to have toughened up agreements reached internationally. An example of this is demonstrated in the phase out of refrigerant hydrochlorofluorocarbons set at 2030 in Copenhagen in December 1992, being tightened in the EC to 2015. That is the date for the ban on HCFC production; but even within that there are tighter sector timescales. From the beginning of 1996 HCFCs are banned in domestic refrigerators and car air conditioners. In 1998 the ban extends to refrigerants in rail air conditioning systems; and for large cold stores and warehouses the ban takes place in the year 2000. Even before these bans take effect and a new crop of replacement refrigerants (hydrofluorcarbons – HFCs) come to maturity, some, like R134a, although ozone free, are coming under pressure to be banned for their global warming potential.

If the drive for quality and safety has received its greatest directional force from industry and government there can be no doubt that the force affecting what we build, what we build it with and how much energy it consumes, receives its greatest power from public opinion. As more and more parts of the construction industry seek to avoid being accused of putting up buildings that send out the wrong environmental impact signals, so they will wish to be seen doing the right thing. At worst clients will have 'the best possible solutions in the circumstances' even in those instances where with commendable honesty 'to make a profit' is one of the deciding circumstances.

Builders have a tremendous record for doing good in the communities they serve and will not be left behind in their efforts to do well environmentally. They have a high class act in place with the Construction Industry Environmental Forum comprising the BRE, CIRIA and BSRIA. Representing membership drawn from clients, designers, constructors, operators, materials producers, service industries, property owners, including central and local government, financial and legal interest, CIRIA and BSRIA published the following documents in 1994:

- CIRIA Environmental Handbooks for Building and Civil Engineering Projects; *Design and Specification* Special Publication 97 [18]; *Construction Phase*, Special Publication 98 [19].
- BSRIA: *Environmental Code of Practice for Buildings and Their Services* [20].

The author acknowledges the support of both organizations in granting their permission to use extracts from their documents.

To complete the circle of primary guidance, in 1995 CIRIA issued Special Publication 120, 'A Client's Guide to Greener Construction' [21].

4.4.2 THE CIRIA HANDBOOKS

The handbooks form 'a useful checklist and inventory of possible impacts, and good practice for organizations considering setting up formal environmental management systems at a corporate or project level'. The composition of the contents and the issues covered are listed in Table 4.13 and Table 4.14.

Table 4.13 CIRIA *Environmental Handbook: Design & Specification,* contents and issues covered

The main sections comprise:
- Agreeing the brief and setting the project environmental policy
- Inception and feasibility
- Primary design choices and scheme design
- Detailed design, working drawings and specification
- Environmental considerations at tendering and contract letting

Issues covered include:
- Policy and forthcoming legislation
- Energy conservation
- Resources, waste minimization and recycling
- Pollution and hazardous substances
- Internal environmental issues
- Land use and conservation issues
- Client and project team commitment

Table 4.14 CIRIA *Environmental Handbook: Construction Phase,* contents and issues covered

The main sections comprise:
- Introduction to the key issues involved
- Tendering
- Project planning and contract letting
- Site set up and management including traffic management
- Demolition and site clearance
- Groundworks
- Foundations
- Structural work for building or civil engineering
- Building envelope
- Mechanical and electrical installations and their interface with civil and building work
- Trades: joinery, painting, plastering, etc.
- Landscaping, reinstatement and habitat restoration or creation
- Site reinstatement, removal of site offices and final clear away
- Handover and guidance on maintenance and records

Issues covered include:
- Policy and forthcoming legislation
- Energy conservation
- Resources, waste minimization and recycling
- Pollution and hazardous substances
- Transport
- Waste disposal
- Client and project team commitment

Table 4.15 Parts of BSRIA environmental CoP related to project lifecycle stages (read with Figure 4.5)

CONCEPT	
Part A	Inception and outline briefing
Part B	Feasibility
Part C	Outline proposals
Part D	System design
Part E	Detail design
Part F/G	Production information
Part H	Tender action
Part J	Pre-constuction
Part K	Construction
Part L	Completion
Part M	Occupation: facilities management Operation and maintenance
Part N	Feedback
Part P	Refurbishment and recommissioning
Part Q	Decommissioning
Part R	Dismantling and disposal
REDEVELOPMENT	

4.4.3 THE BSRIA CoP

Targeted at a similar audience to that of the CIRIA handbooks the BSRIA code proved complex to develop, due to a number of factors, one of which was 'the interdependency of all parties involved in determining the appropriate strategy for a given building'. It defined a role for '*Contractors, Manufacturers and Suppliers* by meeting the requirements in an environmentally sound way; by minimising any wastage, pollution, hazards and risks associated with their products, services and working practices; and by providing occupiers with better training, information and support'.

The code is in parts related to a project's lifecycle stages from concept to redevelopment, reproduced here as Table 4.15. The use of the parts are shown in typical sequences of work stages for traditional, D & B and management procurement methods; see Fig. 4.5.

4.4.4 BREEAM

BREEAM stands for the Building Research Establishment Environmental Assessment Method [22]; and sooner or later the site manager

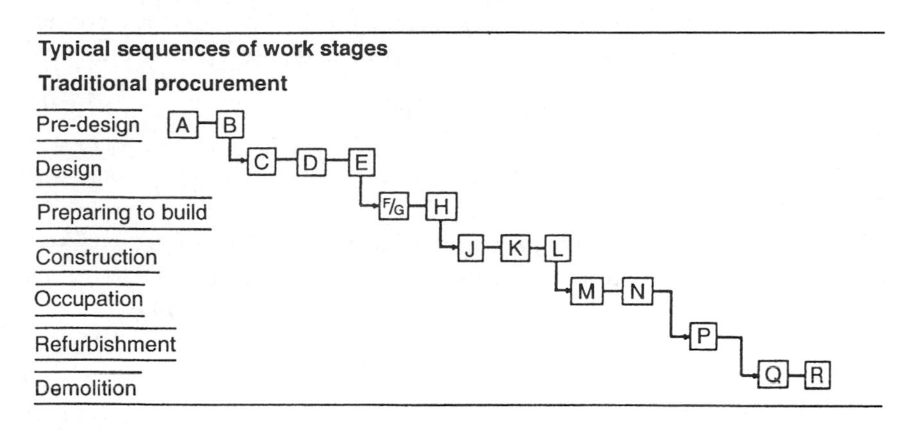

Design–build procurement

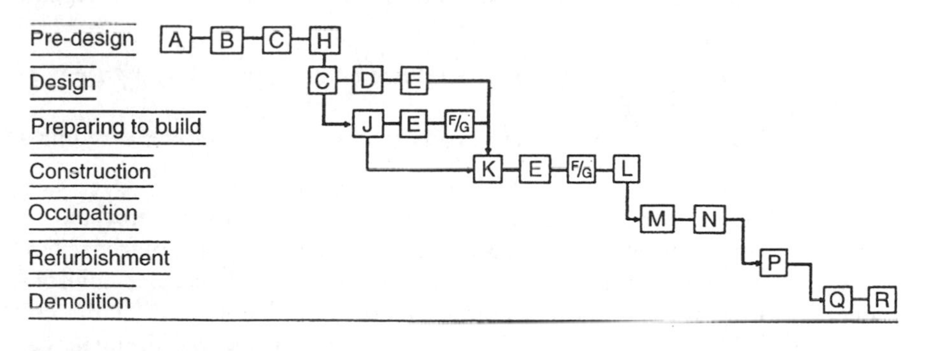

Management procurement

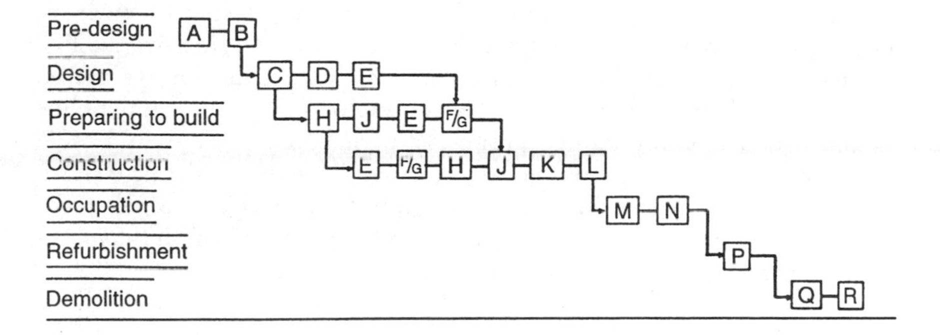

Figure 4.5 Typical sequences of work stages. (Source: BSRIA Environmental Code of Practice.)
Note: Sequences of work stages are not time dependent.

will be involved in a project that has been through this process. The BREEAM scheme is available for a variety of building types. The buildings are awarded credits for environmental features that are considered in the way a building makes an impact under three environmental headings:

- global
- neighbourhood
- internal

The building is scored through recognizing the features of its architecture, structure and BS that interact to provide an assessable rating system. The higher the score the lower the environmental impact.

4.4.5 ROLE OF THE PC

The role of the site manager in relation to the BS contractor will depend on:

1. Whether the contract is being conducted under the PC's own environmental policy or
2. an environmental policy that forms part of the employer's requirements.

In both situations there will be requirements which can be expected to be passed on to the BS contractor. In the case of (1), particularly where it is part of the PC's policy to select subcontractors with similar environmental attitudes to their own, compatibility should not be difficult. For (2), where the client has a higher environment profile requirement for the procurement of buildings, and where the site location makes further demands, there will be a contractual need for project specific environmental plans. These requirements will have already been met by the PC, and BS contractor, the latter through prequalification, tender evaluation and pre-award examination. The BS contractor's environmental policy statement and possibly outline proposals, perhaps no more than a table of contents for his proposed environmental plan, will have been accepted. Alternatively, the PC, as part of the employer's or his own requirements, may have listed the scope and content for the BS contractor's environmental plan compliance at the enquiry stage.

4.4.6 SCOPE AND CONTENT OF BS CONTRACTOR'S ENVIRONMENTAL PLAN

In the absence of any specified requirements from the employer to the BS contractor, via the PC, Table 4.16 may be of help. Where there are specified employer's requirements for an environmental plan the BS

Table 4.16 Suggested scope and content of BS contractor's environmental plan

Company policy statement	–
Project objectives	Summary of project environmental issues and methods of addressing them
Allocation of responsibilities	Assignment of responsibilities and description of duties related to the project's organogram
Review and audit	Programme for regular review of policy and procedure implementation
Control of documents	The preparation, issue and revision of specific method statements, procedures and work instructions related to environmental matters. All method statements, procedures and work instructions should contain a subheading 'Environmental' under which relevant aspects, e.g. waste, pollution, protection, etc. can be addressed
Management of subtraders and specialists	Procedures for vetting methods, procedures, work instructions and reviewing their implementation of the project's environmental requirements
Complaints and contraventions	Procedures for receipt, notification and resolution
Induction and training	For new employees in company environmental policy, and all employees in the specific project arrangements
Records	Of induction and training for all employees. Results of reviews and audits. Details of complaints and contraventions and their resolution
Appendices	Schedules of applicable standing instructions, project procedures, work instructions and related proformas. General should cover: – Control, reduction and disposal of waste – Recycling of packaging material – Control of pollution – Environmental issues in the site offices Specific should cover the particular employer's requirements applicable to the project, not covered by general

contractor's proposals could be checked by the site manager against those of his company, which by this time can be expected to have been approved. Where requirements to be complied with affect both contractors those from the BS firm should at least match those of the PC. If the

PC's environmental plan has been drawn up using the CIRIA handbook, the site manager may reasonably expect the BS contractor's plan to comply with guidance in these documents:

- Stage C.3, *Site Set up and Management,* which covers site specific planning plus general guidance on environmental management of the site on a day-to-day basis. Sub Stage C.3.9, 'Environment in the site offices' is particularly relevant.
- Stage C.9, *Mechanical and Electrical Installations and Their Interface with Civil and Building Work* 'covers briefly the interface between the different disciplines and the special steps they need to take to ensure that their independent actions do not jeopardise the environmental performance of the project.' Sub-stages C.9.1, 'Legislation and policy' and C.9.2, 'Co-ordination between structural, civil and service contractors and the use of BSRIA environmental code of practice' are also relevant.

4.4.7 MANAGEMENT OF BS CONTRACTOR'S ENVIRONMENTAL PLANS

Table 4.17 has been taken from the CIRIA handbook, 'Construction phase', C.9.2; it provides a six-point framework for site management

Table 4.17 Good practice for site management between BS and other contractors (Source: CIRIA *Environmental Handbook* Publication No. 98.)

Issue C.9.2 Good practice

Site management should, when dealing with interfaces between the main civil or building contractors and the services contractors:

- ensure all parties are aware of the project's environmental policy and of each other's corporate environmental policies;
- set up regular liaison between each of the contractors' environmental managers such that they ensure no conflict in environmental terms between the various contractors' actions;
- ensure all contractors' environmental managers are aware of the strategy and guidance outline in BSRIA's Environmental Code of Practice for Buildings and Their Services in assisting contractors to implement coordinated and practical, environmentally responsible procedures;
- ensure environmental plans drawn up by the individual contractors following the guidance in this *Handbook* are complementary;
- consider the preparation and distribution of contract 'trees' showing all the contractual and environmental inter-relationships on the project;
- consider the implementation of punitive measures against any contractors and subcontractors for non-compliance with the project environmental policy.

Table 4.18 (a) On-site good practice and (b) waste minimization (Source: BSRIA environmental CoP Part K, construction notes.)

Good neighbourliness
(a) On-site good practice
  Provide time schedules to local inhabitants affected
  Avoid noise, on-site burning, mud on raod, anti-social hours
  Avoid obtrusive lighting or security procedures
  Protect trees and local ecology
  Follow health and safety procedures:
  – good ventilation
  – prevention of dust
  – orderly activity
  – simple precautions with regard to moving equipment and electrical items
  Use environmentally friendly transport
  Avoid pollution escape to air, water or land
  Avoid on-site chemical treatment
(b) On-site waste minimization
  Care of materials/goods
  Minimize wastage including water and environmental impact
  Use recycled materials where fit for purpose
  Salvage and recycle waste building materials and packaging
  Consider prefabrication to avoid waste or at least predetermined sizes, quantities and loads

'when dealing with interfaces between the main civil or building contractors and the services contractors'.

For the site manager to ensure that the BS contractor is aware of the strategy and guidance contained in BSRIA's *Environmental Code of Practice for Buildings and Their Services*, means that he too must have knowledge of its contents. This is not such an onerous task, as can be seen from Table 4.18 BSRIA CoP Part K, construction notes: 'On site good practice, and on site waste minimisation'. The level of guidance is very basic and should allow the site manager to return to the comfort of the CIRIA handbook where in stage C.3.9, 'Site set up and management', more extensive advice will be found; see Table 4.19.

This comfort of a common approach to all the general matters that will concern every site should allow the site manager to concentrate on understanding those environmental issues which are particular to the construction and commissioning of building services on his project.

Environmental management should be an item on the agenda of every site progress meeting. Its position in the order of play should be near the top so that it is never submerged by pressures of time and cost that ebb and flow with varying intensity. The treatment of the site's environmental issued should never be relegated to the 'if we've got time, we'll

Table 4.19 Site set up and management issues covered in CIRIA *Environmental Handbook: Construction*, Stage C.3

C.3.1.	Legislation and policy
C.3.2.	Positioning, layout and planning of the site compound
C.3.3	Relations with neighbours and the local public
C.3.4	Implementing green management plans on site
C.3.5	Protection of sensitive areas of the site
C.3.6	Implementing pollution control strategies
C.3.7	Traffic management
C.3.8	Environmental impact of temporary works
C.3.9	Environment in the site offices
C.3.10	Special consideration for specific civil engineering projects

do something about it' attitude. Those of relaxed, unwary or dismissive approach are reminded of their liabilities under the Environmental Protection Act 1990 [23], its subsidiary regulations, and EC directives, the main legislation of which is listed in the CIRIA handbooks.

References

[1] BS5750 Quality Systems.
[2] BS4778 Part I 1987 Quality Vocabulary Standard.
[3] The National Association of Lift Makers (1994) *Principles of Planning and Programming a Lift Installation*, London.
[4] *The Construction (Design and Management) Regulations 1994*, HMSO, London.
[5] The Health & Safety Executive (1994) *Essentials of Health & Safety at Work* HMSO, London.
[6] Waterman, Lawrence (1993) 'What is 'Risk Assessment' anyway?' *Health and Safety at Work*, January.
[7] The Health & Safety Commission (1995) *Approved Code of Practice Managing Construction for Health and Safety*, London.
[8] The Health and Safety Commission (1995) *Designing for Health and Safety in Construction*, London.
[9] CIBSE (1995) *Health, Safety and Welfare: Guidance for Building Services Engineers*, TM 20, London.
[10] The Heating & Ventilating Joint Safety Committee (1994) *Site Safety: A Guide to Legal Responsibilities for Health, Safety and Welfare*, London.
[11] The Heating & Ventilating Contractors' Association (1994) *Risk Assessment Manual*, London.
[12] Hascom Network Ltd (1993) *The ECA Risk Assessment Manual*, Hascom, Southampton.
[13] The Control of Substances Hazardous to Health Regulations (COSHH) 1994, HMSO London.

[14] The HVCA (1990, 1991) *The HVCA COSHH Manual*, Vols. 1 & 2, London.
[15] HVJSC (1994) *H & V Safety Guide*, London.
[16] Latham, Sir Michael (1994) *Constructing the Team*, HMSO, London.
[17] Building Employers' Confederation (1984) *A Guidance Note on Safety Responsibilities for Sub-Contractors on Site*, London.
[18] CIRIA (1994) *Environmental Handbook for Building and Civil Engineering Projects, Design & Specification*, Special Publication 97, London.
[19] CIRIA (1994) *Environmental Handbook for Building and Civil Engineering Projects, Construction Phase*, Special Publication 98, London.
[20] BSRIA (1994) *Environmental Code of Practice for Buildings and Their Services*, Bracknell.
[21] CIRIA, *A Client's Guide to Greener Construction*, Special Publication 120, London.
[22] BRE (various years) *BREEAM: Building Research Establishment Environmental Assessment Method.*
[23] The Environmental Protection Act 1990.

Planning and programming 5

5.1 General

Planning is strategic, while programming is the organization of events by which strategy will be implemented. For a small, simple, single building with an uncomplicated internal layout, strategy is straightforward. The programming activities automatically come to mind. They are foundations, frame, envelope, internal divisions, finishes and building services. Even the degree of overlap of one stage starting before another finishes, and how early can we start the building services, are perhaps not difficult to solve. On jobs of greater constructional and building services complexity the overall planning and programming is developed and tested through a number of substrategies implemented through logical sequences of activities. But we are running ahead of ourselves, for once again we must consider the form of the contractual arrangement.

Traditional JCT lump sum tendering is still the most favoured form of contract; sharing a close affinity is D & B. Some of the better lessons learned from managed contracts are being applied to the enquiries for these lump sum type tenders, in the demand for more sophisticated programming. The managed contracts benefited from tendering contractors seeking an input for their bids from BS contractors and specialists. This help from the engineering industry was proffered on the basis of securing a place on the works or trade contractor bid list. The traditional tendering contractor has not been so fortunate. Tendering periods are short, shackling the planning manager. Even if he wants to seek advice from industry, who does he speak to? The BS contractor to be used in the main contractor's bid is probably not known to the planner until, if he is lucky, 24 hours before the main tender is due in. If, as is not unusual, proposal documentation forms part of the contractor's tender submission, and a not unreasonable level of programming information has been requested, e.g. overall, construction and finishes programmes, then the planner's work will have been finished long before he knows the name of the services contractor. In these most common of circumstances the planning manager is left to his own skill and resources.

Many planning managers have quite rightly risen to seniority on the back of good performances on MC and CM projects on which they controlled the flow of all information other than financial. Supported by on-site staff and banks of computers they have juggled numerous works/

trade contractors and all the information they generate and request, into logical predictions, reasonably accurate progress records, and completion date forecasts. Promoted with the friction generated warmth of major projects behind them some planning managers are submerged by the sheer volume of traditional type work that lacks the excitement of the major project. They have to handle this workload without the back up they enjoyed from construction and building services managers on site. Of course this is not the only reason why services are ill considered in the planning and programming of traditional building work. The intensity and complexity (and these usually infer risk) from building services are not appreciated. Fear of failure inhibits 'traditional' contract builders from showing BS activities too closely coordinated with construction work. Should they not be able to make an area, floor or ceiling, or other work face available to the BS contractor at a programmed time the builder is afraid of creating a 'claim for delay' situation. These are the key reasons why we have too many jobs with overambitious, simplistic programmes that are drawn to show building services as a stage whose start and finish is related to other more easily defined construction stages. But we know that BS are really not like that. Plant, equipment and components are combined to form systems which are threaded through, fixed to and placed on the structure and fabric of the other work. They are dependent on those works and as every apprentice or trainee who was sent to the stores for some 'sky hooks' or 'box of assorted holes' has found out, cannot be fixed before the building work exists. Even then some of that building work has to have particular activities carried out to facilitate the construction of BS. These works are known as 'builder's work in connection' (BWIC) with building services. Information is required for that preparatory BWIC, but it is far from being the only piece of information that is required on site, before BS work can commence.

5.2 Delivery of information

5.2.1 GENERAL

We commenced this information delivery process in the previous chapter by dealing with the preparation of quality, safety and environmental plans and will continue with planning for BWIC, working drawings, method statements, sample approval, test and inspection plans and procurement schedules. The programming of the delivery of these provides the site manager with a tool for monitoring the construction progress. The site manager's views of the BS contractor's performance at information delivery stage will be his marker for how carefully he looks at the early installation activities.

The importance of understanding what information has to be planned before construction should be allowed to commence on site can be seen from Table 3.1. Out of the 18 items 6 are dealt with here. No work should commence without knowing what has to be done, where, in what way and by whom. The status of pre-installation information provides early assurance of the BS contractor's intention that the work will be to the specified requirements, which will be met safely and with minimum impact on the environment – see Chapter 4.

The wise site manager will not leave the delivery of pre-construction information to hoped for action by the BS contractor on the filed away minutes of the pre-award/management strategy meeting (see Table 3.2). Nor will he leave it to the pre-start site meeting. For if he does, regrettably, he will find that little or inadequate progress has been made by the BS contractor. Depending on his judgement of the size and complexity of the building services the site manager should call for an information production planning meeting between 4 and 12 weeks before start on site. At that meeting the site manager should go over the scope and content of the information that is required and agree an early date for the issue of an information release programme. Table 5.1 lists the main areas to be covered. Prior to the meeting the site manager should refresh his memory of the BS content by flipping through the particular requirements of the specification and turning over the BS general arrangement (GA) drawings. It is also worth remembering that the BS contractor will not be able to provide all the information prior to construction. A number of influences, depending on how many specialist sublet contracts there are, will determine the availability of information. But planning for the delivery of information in this way should ensure that the orders for these specialist sublets are placed in time. On large jobs the site manager may be supported by a resident or visiting planning manager and it is not uncommon for this manager to control the receipt, logging and distribution of project documentation. Numerically, the greater number of jobs are those where the office based planning manager hands tender proposal programmes to the site manager and leaves the rest to him.

Table 5.1 Agenda for an information production planning meeting

Delivery of information

1. Quality, safety and environmental plans
2. BWIC drawings
3. Working drawings
4. Method statements
5. Sample approvals
6. Construction test and inspection plans (including offsite manufacture)
7. Procurement schedules

5.2.2 BWIC BUILDING SERVICES

Table 5.2 will certainly need the site manager's agreement, possibly in consultation with the general foreman on the subject of holes and chases in brick and blockwork. It is bad management to ask a BS contractor to produce drawings for these, only to be told at the work face, 'We don't want drawings – we want you to mark it out.' On MC and CM projects where many of the BWIC holes, chases and fixings are down to the BS contractor, it is sensible for the site manager to agree the principles and detail of this work.

Penetrations through the structural frame, walls and floors may have been indicated on the structural engineer's drawings. Quite often the structural engineer is only interested in holes over say, 150 mm × 150 mm, and the site manager should not be surprised when the services contractor asks for a number of smaller holes to be provided that are not shown on the structural engineer's drawings. The ingress and egress of utilities may also require the building in of multi-way pitch fibre or earthenware ducts, and steel/cast iron puddle flanges, usually through the walls of basements or foundations.

A word about secondary steelwork. This can be required for trimming roof ventilation openings, plant room air intakes and discharges, and supports for services suspended between the spans of the building's

Table 5.2 General range of builders' work for building services

Bases for
- Plant e.g. boilers, pumps and transformers
- Equipment, e.g. control panels.

Holes (and chases)
- Walls
- Floors
- Ceilings

} Location

- Cut (mark out)
- Form (in shuttering)
- Leave (in brick and blockwork)
- Prefabricate (in steelwork).

} Method

Fixing
- Built in, e.g. unistrut
- Drilled
- Fired
- Clamped or welded (to steelwork).

Secondary steelwork for
- Supports
- Access, e.g. cat ladders
- Protection, e.g. guardrails.

structural frame. The connection the primary support can take the form of building in, bolting, cleating and welding.

The BWIC provisions for lightning protection due to its earthing function and association with the groundworks requires the provision of early information, sometimes immediately followed by the provision of enabling works in the form of earth rod pits. This can make it good sense to have the lightning protection specialist employed direct by the builder rather than waiting for an appointment via the electrical contractor.

In addition to these general considerations there are a whole host of particular requirements of which templates for casting in holding-down bolts for free standing or guyed flues, is but one. It is not unusual for the BS contractor to have to provide BWIC information soon after his own appointment but before orders can be placed and information sought from his specialist subtraders. The BS contractor is at risk in providing such information when the 'intelligent guess' derived from catalogue information happens to turn out as incorrect when the manufacturer's 'certified' drawings come along. The PC can do much to help himself and mitigate this potential cause of conflict by recognizing the complexity of BS and placing the order as quickly as possible. By doing this the contractor is providing sufficient time for the BWIC and all other pre-construction information to be properly managed.

The site manager is recommended to study the interface between *in situ* reinforced concrete and electrical services in areas such as staircases. Here surface mounted light fittings may be fed by wires running in embedded conduits. It is not general industry practice for the designer or installer to produce conduit layouts. They are, however, necessary where they are embedded in concrete so that the shuttering formwork can be designed and the prefabricated conduits fixed (with protected ends) by the formwork carpenters.

At this stage the site manager should take the opportunity to increase his understanding of whose responsibility, and by what method, holes will be made good following the installation of building services. Holes are required to be made good to prevent air leakage, heat loss or gain, and as magnetic field barriers, but most importantly for fire stopping and acoustical reasons. BS penetrations occurring through fire compartmentation walls and floors must be made good to the mandatory requirements of the Building Regulations. In cases where the designer has not detailed methods the contractors must determine the ways of achieving fire stopping. Table 5.3 gives the considerations for selecting the appropriate method. Table 5.4 (a) and (b) list some of dry and wet methods in use. The planning and programming of BWIC drawings will be similar to that described next, for working drawings.

Table 5.3 Fire stopping – selection considerations

Performance, measured as stated in BS 476
Effectiveness
Durability
Adaptability
Compatibility with surroundings
Convenience of application
Ease of removal
Cost

Table 5.4 Fire stopping: (a) dry methods; (b) wet methods

(a) Dry methods
Cable/pipe transits
For cables and pipes of mixed size passing through solid floors/partitions of high fire rating
Robust but adaptable
Solid intumescent materials
Expands to fill gaps at trigger temperature
Will not stop cool smoke
Intumescent bags
Fire resistant bags containing intumescent granules
Used to fill medium size holes/voids
Easily removed and replaced
(b) Wet methods
Weak mix mortar
For large cables passing through solid floor/partitions which are unlikely to be disturbed
Intumescent plaster
Used to fill medium size gaps
Intumescent mastic
Used to fill small, difficult gaps

5.2.3 WORKING DRAWINGS

The submission of a list of working drawings to be produced by the BS contractor should be called for. The list can be offered to the design engineer for comment on its likely adequacy. This has a twofold benefit. The site manager gets an expert's opinion on the adequacy of the numbers of drawings intended to be produced and the designer is made aware of his future commentary/approval work load.

The working drawing list will only be provisional and the BS contractor should not be held to providing exactly the number of what are, after all, his first thoughts for general arrangement and main coordination details. In reality more drawings will be required as specialists are brought on board by the BS contractor. These will produce function diagrams, schematics, wiring diagrams, shop drawings and manufacturers' certified drawings, etc., etc. When calling for the preparation of a working drawing list, which takes very little time produce, the site manager should ask for the list to be supported by a programme of drawing production. This is a not too difficult task, produced in hours rather than days. So far the BS contractor has had no difficulty in meeting the site manager's request. The next question is the real test. Ask the BS contractor to turn the production programme into one which shows the assessed man days for the preparation of each drawing and name the resource who will produce it. Then we are getting down to the creation of a proper monitoring tool. Any production slippage can be addressed and doubts about the availability of the named resources checked by a head count in the contractor's office. Not only should heads be counted but also CAD terminals, their operators and level of shift working. Figure 5.1, typical programme for production of drawings, was produced by the author for BSRIA TN17/92 *Design Information Flow,* the principles of which are just as applicable for working and BWIC drawings.

The BS contractor should be informed that the drawing production programme must have sufficient 'front end' allowance before commencement on site for the drawings to pass through an approval stage.

If no prespecified approval circuit or status protocol has been advised to the BS contractor, these details must be negotiated, e.g. drawings will be approved:

- A: no comment;
- B: approved subject to comments;
- C: redraw.

Before agreeing a procedure first check with the designer what his role will be in the process and determine whether he will be giving approval or comment. Talking to the BS designer will usually elicit if there is a DT procedure in which working drawings are issued and comments collated via the lead consultant or direct with each design discipline. Control via the lead consultant will add time to the approval process.

Finally before leaving the subject of working drawings the site manager should call up the BS contractor's documentation for drawing control, i.e. register of issue and approval status.

5.2.4 METHOD STATEMENTS

At this point in the production information meeting proceedings the BS contractor in response to the site manager's request for method state-

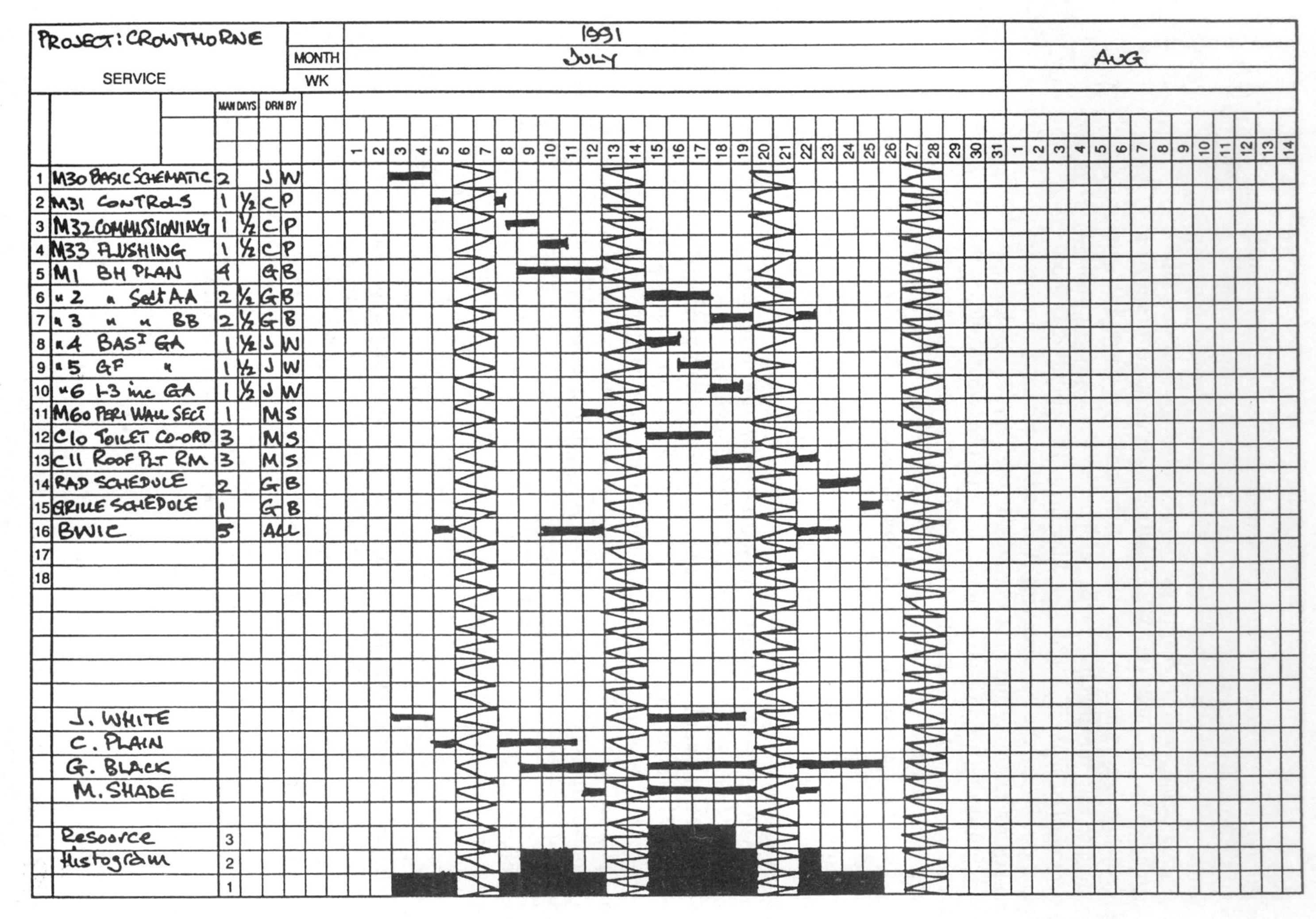

Figure 5.1 Typical programme for production of drawings. (Source: BSRIA TN 17/92, *Design Information Flow*.)

ments should be saying something like, 'I will extract the list of Method Statements from the Quality and Safety Plans I am submitting to you next week.' The expected list may be similar to Table 5.5. Note that even the general list contains a high number of sublets. These are termed 'Traditional' sublets for which, on uncomplicated jobs, the BS contractor will have no difficulty in integrating their activities into methods of working prior to the subtrader's appointment. On complex projects the BS contractor may wish to revise his preliminary methods after the appointment of the traditional sublets. The date for the programme showing the release of these statements to the PC would be agreed.

The privatized regional gas, water and electricity companies, with British Telecom, Mercury and others in the utilities business, provide an element of BS to a project, as contractors. Regrettably, the previously

Table 5.5 Coverage of (a) general method statements and (b) method statements for typical specialists and subtraders

(a) General	
Mechanical (including plant erection)	
Piping	
Ducting	'Traditional sublets'
Insulation	
Controls	
Public health (including plant erection)	
Plumbing (soil, waste and vent piping, sanitary fittings)	
Water services (HWS may be in mechanical)	
Sprinklers, hosereels, dry fire riser	
Lifts and escalators	
Utilities	
Electrical (including plant erection)	
Power	
Lighting	
Containment systems (for specialist sublets)	
(b) Specialist	
Lighting protection (E)	
Substation and HV switchgear (E)	
Security (E)	
Data (E)	
Telecoms (E)	
Fire detection and prevention (E)	
Kitchen and cold rooms (M)	
Generators (E)	
UPS (E)	
BMS (M) or (E)	
Commissioning management (M)	

public supply authorities still act with autocracy, and present the site managers with great difficulty in obtaining programmes, method statements and all the usual information normally provided by a BS contractor. So difficult is this process of extracting the properly documented, intended and delivered performance from the utilities that they can jeopardize a contractor's registration to BS5750.

The site manager should remind all of the BS contractors that he will expect to see the H & S risks identified and assessed in the method statements, and proposals included for safe working practices.

The programmes for method statements release should show the latest date by which the BS contractor will be appointing his specialists and traditional sublets. It is prudent to agree a 'turnaround' time for the site manager's commentary on the method statements. This should allow for any requirement for the statements to be subject to the comments of the designers' and PC's H & S adviser, both of whom may be off site. In establishing a turnaround timescale the site manager should make allowance for the need of any particular area of work that may require the opinion of his H & S adviser. Such method statements should be identified and given an agreed longer time. Coverage on how and what to comment on will be found in Chapter 7, as we are only concerning ourselves here with the timely delivery of information.

5.2.5 APPROVAL OF SAMPLES

At the information production planning meeting the BS contractor should be requested to produce a list of the specified requirements for the approval of materials, samples, mockups, trial site assemblies and workmanship. Designers' reasons for specifying these requirements are for 'setting the standard', some aspects of which will not only be 'fit and tolerance' but aesthetic and therefore subjective. Another reason is innovation, where perhaps well-known materials are being applied in a new way, or a new product is being applied. True prototyping may not be obvious although many subscribe to the theory that every building project is a one off. Whatever the particular reasons for specifying the provision of samples some definitions may help to understand the importance of planning for their delivery.

Material – the substance of which a thing is made or composed

Usually the material samples are concerned with what the ducting, piping and electrical distribution systems are composed of. Therefore materials here may include components such as fixing assemblies made up of one or more materials. This may also apply to the finish of a product such as a grille, luminaire or control panel colour, etc., etc.

Samples

Most frequently applied to terminals, i.e. the fixtures at the end of distribution systems in spaces of permanent or transient occupancy, e.g. luminaires, grilles, switchplates and socket outlets. They could apply to components built into or at intersections in services systems, e.g. fire damper, distribution board or earth leakage device.

Mock ups

Care is needed here, as these may either be a fully fitted out part of the BS installation integrated with the structural and building fabric elements such as a perimeter module of an office, or may be purely aesthetic. The latter gives the impression to the viewer that the hidden workings are included, for example the same office perimeter but with empty skirting trunking and builder's work heater casing with air discharge grille. It is not unknown for approved aesthetic mock ups to have insufficient capacity to accommodate the essential contents of their workings, e.g. the fan coil will not fit or all cables go into the trunking. Full-scale mock ups are best.

Trial site assemblies

Normally these are concerned with setting the standard, for example in the first toilet constructed in a building the DT may wish to approve the arrangement of waste and water piping below a range of wash handbasins. When approved this would be the 'marker' for all other toilets of similar layout.

Workmanship - the degree of art or skill exhibited in the finished product

A standard setting exercise to be repeated wherever the sample example occurs throughout the installation, e.g. the making of an on-site joint between two offsite manufactured prefinished 'colour coat' sections of ductwork.

Having got the BS contractor to list all of the above, which we will refer to generically as samples, the exercise should be extended to create a 'samples file' of all the specification clauses that refer to them. On projects where the design engineer's specification is available on disk and the BS contractor has computers to read it, a search of the wording along the lines of 'the contractor shall provide samples – materials – mock ups – trial site assemblies' etc. makes the extraction task easier.

The site/planning manager should insist on a programme for the delivery of samples as there may be some reticence on the part of the BS contractor to do so. This is possibly due to a clash of financial interest within the BS contractor's organization. It may be found that the BS estimator has included prices where it was the manufacturer's intention to supply samples after receipt of the order. Many samples are required to be approved early before the BS contractor can place the full order with their source. In these circumstances the services firm finds that samples are being treated by the suppliers as 'specials' at increased cost. Financial tactics ensue, samples arrive late and, with the risk of rejection attendant upon the approval process, can cause delay not only to the BS contractor who may have proceeded at 'risk', but with following trades.

The status of samples must be controlled. Figures 5.2(a) and (b) both from Drake and Scull Engineering are examples of sample control. The simpler general sample submission form 5.2(a) was used where the BS contractor was able to use a form of his own creation. Figure 5.2(b) was prepared to meet more stringent requirements and was supported by a 10-item dossier. The site manager should acquaint himself of the contracts requirements for labelling, storing and access to approved samples. It is important to know if the specification allows for approved samples to be built into the permanent works.

Unfortunately a great number of building services specifications still contain an onerous catchall worded, e.g.:

> The BS Contractor shall include in his Tender for the cost of submitting to the Engineer for approval such samples of workmanship, materials and equipment as the Engineer shall require.

Financially punitive, the late call for samples of work in progress may at worst become disruptive as the BS contractor proceeds at 'risk' while the designer deliberates. Sympathetic support is the least that the site manager can offer his contractor in these circumstances. Any pushing by the site manager for the designer's or client's speedy approval would do no harm to the working relationship.

5.2.6 TEST AND INSPECTION PLANS

As for samples so it is with test and inspection plans. The BS contractor should be asked to produce a two part listing, the first of which should be the provision of these plans as specified requirements, and the second should cover the services firm's own proposals. Remind the BS contractor to include all specified or proposed construction testing. Many contractors believe that test and inspection plans only apply at the commissioning stage. This is not so. Taken simply, all BS distribution ducting, piping, electrical trunking, conduit, cabling and busbars are containment systems. As they are built they must be tested progressively to ensure that they are capable of containing what flows through them, whether it is air, water, gas, sewage or electricity, etc. The BS

SAMPLE SUBMISSION

TO:................................. SUBMISSION No:..................

.... DATE:............................

PROJECT:............................ SERVICE:.........................

....

APPROVAL OF THE FOLLOWING SAMPLE IS REQUIRED BY:-......................

EQUIPMENT:.......................... MAKE: ....

DESCRIPTION:.......................

SPECIFICATION REFERENCES: DRAWING REFERENCES:

....

....

DRAKE & SCULL NOTES

SUBMITTED BY ....

SIGNED .... Date ....

To be completed by the Main Contractor/Consultant/Client

ACCEPTABLE/REJECT

SIGNED:........................... DATE ....

COMMENTS (if applicable)

Form:- NQ.35b

Figure 5.2(a) Sample submission form from Drake & Scull Engineering Ltd.

TECHNICAL SUBMISSION

TO:- .... SUBMISSION NO.:-............

.... DATE:-.........../.../.....

.... SERVICE:-...................

PROJECT:-..

....

APPROVAL OF THE FOLLOWING EQUIPMENT IS REQUIRED BY:- ....

EQUIPMENT:-.. MAKE:-..............................

DESCRIPTION:-...

SPECIFICATION REFERENCES:-.......................... DRAWING REFERENCES..................

....

ATTACHED DETAIL DOCUMENTS

Identification reference of each supporting document is to be entered in the space against the headings (i) to (ix) as appropriate.

(i)	EQUIPMENT SPECIFYING SCHEDULE	(i)	....
(ii)	MANUFACTURES DATA	(ii)	....
(iii)	DESIGN CHECK CALCULATIONS	(iii)	....
(iv)	CERTIFIED PERFORMANCE LEVELS	(iv)	....
(v)	QA / QC DOCUMENTS	(v)	....
(vi)	BUILDERS WORK REQUIREMENTS	(vi)	....
(vii)	ASSEMBLY & INSTALLATION DETAIL	(vii)	....
(viii)	O & M INSTRUCTIONS	(viii)	....
(ix)	LIST OF RECOMMENDED SPARES	(ix)	....

NB:- In submitting this selected equipment we have not reviewed the total system and its performance against the design intent.

COPIES TO:- ARCHITECT
CONSULTANT
MAIN CONTRACTOR
FILE
OTHERS

SIGNED:-...............................

For:- DRAKE & SCULL ENGINEERING

ORIGINATING OFFICE:-....................

....

Form:- TS.51b

Figure 5.2(b) Technical submission form from Drake & Scull Engineering Ltd.

contractor's test and inspection plans must include construction testing and a programme agreed for the release of procedures. The procedures may be considered as submethod statements to be cross-referenced to the main method statements. Inspection procedures will be given a fuller treatment in Chapter 7.

5.2.7 PROCUREMENT SCHEDULES

Not usually specified for circulation to the DT the importance of these to the site manager should not be underestimated. At the pre-production planning meeting the BS contractor should be requested to produce schedules to indicate his intentions for placing on order specialist and sublet trades, major plant, and high volume items such as radiators, grilles, AHUs, fan coil units, luminaires, etc., etc. Proper scheduling can tell us so much about the job – but more of this in Chapter 6.

5.2.8 OFFSITE MANUFACTURE

Combining matters pertinent to procurement schedules and test and inspection plans offsite manufacture requires the BS contractor to conduct another trawl through the specified requirements. The object of this is to produce a list of plant and equipment that calls for arrangements to be made for production stage inspection and test witnessing by the designers and or client. From the extracted information the BS contractor can produce a programme which may vary from simple to complex, as appropriate. Table 5.6 is a typical listing of plant and equipment subject to offsite inspection and testing.

Table 5.6 Typical listing of plant and equipment subject to offsite inspection and testing

Plant
Chillers
AHUs
HV switchgear
Generator
Lifts
Equipment
Fan coil units
Control panels
Prefabrication
Plant rooms
Risers (and riser frames)
Toilet modules

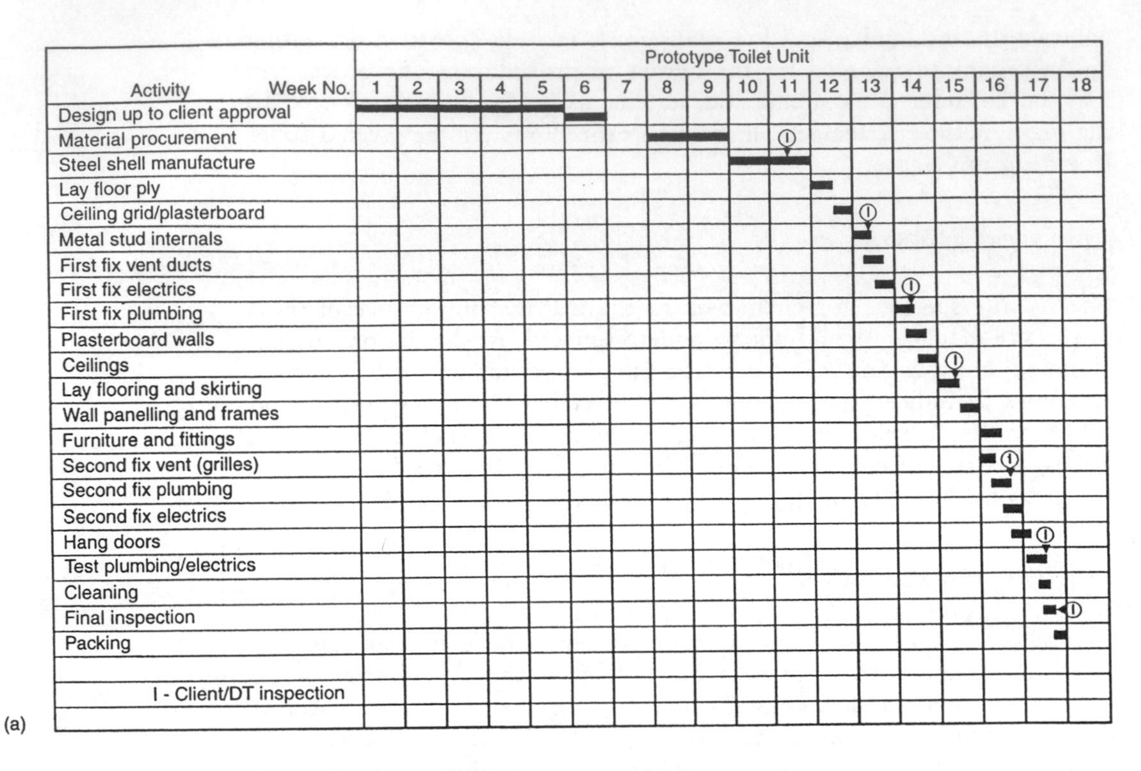

Figure 5.3(a) Prefabricated toilet module prototype production programme.

Out of the monitoring programmes of the BS contractor will come the specific production programmes, with inspection and test dates indicated. Offsite prefabrication of major subelements such as plant rooms, riser and toilet modules may be subject to more continuous inspection. Figures 5.3(a) and (b) show production programmes for a prefabricated toilet module prototype and its subsequent production run for 14 units.

Whereas a few days at infrequent intervals to inspect pieces of plant and equipment do not stretch site resources, the need to inspect large numbers on a regular basis does. The availability of the BS contractor's resources to carry out these actions should be discussed at the production programming stage.

5.2.9 SUMMARY

At this point it is hoped that the site manager can see the benefit of sitting down with the BS contractor and agreeing what pre-construction

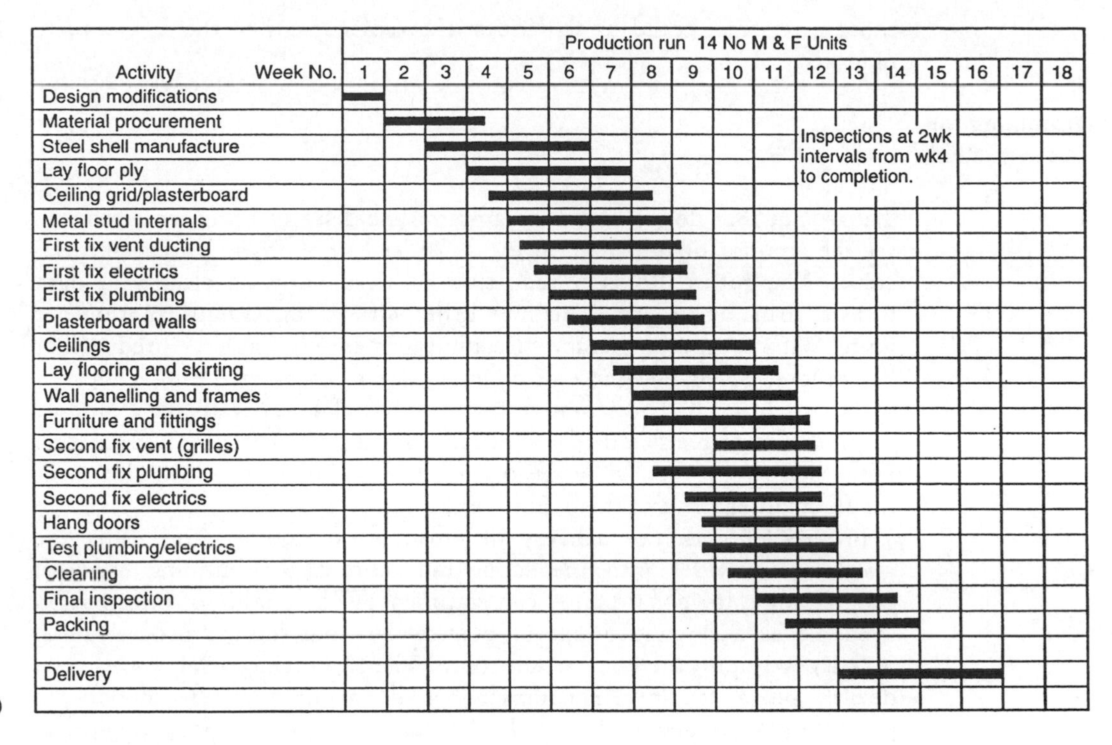

Figure 5.3(b) Prefabricated toilet module production programme for 14 no units.

information is required, when it will be released, to whom and in what format. As a result, the listings of information to be prepared and programmes for its release produced by the BS contractor can be used by the site manager for monitoring the production performance, the expectations for which have been clearly defined. If the BS contractor goes on to fulfil delivery to the promised dates, then the foundations for a good working relationship will have been laid. Should the early promises to perform not have been fulfilled the site manager has to find out the underlying reasons. There is always a tremendous amount of work to be done on starting up any major contract or building services subcontract and the site manager will be aware of this. He will need to make judgements as to whether the BS contractor is underresourced or has good resources but is a poor manager, making promises he knows he cannot perform. We are entering an area where the site manager needs to proceed carefully and he should discuss possible courses of action with his contracts manager. Perhaps soundings can be taken from the last job led by the BS contractor's site manager. The feedback may be bad, but as we are at an early stage in the project, the best solution, though extreme, could be to request the replacement of the services site manager.

5.3 Programming for construction

5.3.1 GENERAL

The output that flows from programming the delivery of information will see the amount of information in use on site take off on an exponential curve. The future holds a reduction in hard copy as electronic data transfer firms up its grip. This may bring with it not less information but more. With computer literacy spreading is it unreasonable to predict that before long we shall see interdisciplinary site meetings planning a cycle of work, where each representative calls up his relative details on a laptop? After all design teams do this with systems like the Autocad Workcentre. Perhaps we may then truly see drawings and information 'on site' with the foreman at the workface. But we digress.

In parallel with the delivery of information exercise and using the insight it provides into the scope and content of building services, programming for construction can commence. The quality of information available as a starting point for post-tender construction programming will depend upon the contractual route. The site manager left to his own devices on a traditional contract has a lot of work to do if the programmes handed to him only list services activities as a single bar chart line for M & E or at best first fix, second fix and final connections. The site manager supported on a management contract by a resident planning manager, whose detailed planning in resolving some difficult services aspects scored points in the evaluation of fee bid proposals, is far better served. Reducing these extremes to the common denominator, that there is always room for improvement, we will look at the substance of what has to be programmed rather than the programme methods. Arrow or precedence diagrams, line of balance, critical path analysis (CPA), etc., etc., and hierarchies of programme types such as 'six week look ahead', short-term and recovery programme are taken as being understood in their relevance and application. Once again we will concentrate on what is different about building services, the range of activities, sequence, geographical locations, e.g. plant areas, and resourcing. Differentiation of major and minor interfaces with other works and the limitations of building services dimensional modules with those of the building structure and fabric will be covered.

5.3.2 IDENTIFY SERVICES

Site and planning managers are trained in depth and understand the variety of structural systems, building envelopes, internal divisions, floor and ceiling types that create a building's footprint and what stands upon it. They are less happy in their understanding of building services that create internal space environments, detect and protect for fire and security. They seem reasonably at home with the public health services, after all underslab drainage, rainwater disposal and sometimes the above ground sanitary plumbing is 'billed work'. They struggle with the BMS, Data (i.e. information technology) and Telecoms, so what exactly does 'Identify services' mean? Table 5.7 could be subtitled 'Know the Job'!, and is our starting point. When going down the list refer back to Table 1.1. Using the job specification tick off from the scope, content and description of services section the range of services on your job. Using the same parts of the specification it should be possible to extend the generic list to the right with system types. Some designers confusingly mix these up, e.g. a specification may say 'the sales area shall be air conditioned (no mention of system type) and provided with high frequency fluorescent lighting (type of lighting is given)'. Next get a feel for how many systems, e.g. each air handling unit (AHU) will be serving a separate ventilation or air conditioning system; 6 no. AHUs = 6 no. HVAC systems. Look for descriptions of services zoning and what services serve which areas. To find out what goes where? Look in the specification for descriptions as to where the BS plant is located, along what routes they are distributed, and where terminals are positioned. In all of this search for information be armed with the BS tender drawings which should confirm interpretation of the written word. Now reverse the process and taking the drawings start to get a feel for the BS interfaces with structure and fabric.

Table 5.7 Prerequisite knowledge for programming BS construction

---

KNOW THE JOB!!
Identify the services
List generic systems
List system types – description?
    System numbers e.g. how many AHUs?
    What do they serve – room name/space function
Find out what goes where
    Plant – local/central
    Distribution routes
    Terminals, e.g. surface/recessed
Look for the interfaces with structure and fabric, e.g. passing through, fixed to/against, covered by

---

5.3.3 INTERFACES WITH STRUCTURE AND FABRIC

The way in which BS interface with fabric and structure are treated as, major, minor, modular – at walls, floors and ceilings, and finally specials. Only through knowledge of where and how BS relate to the building structure and fabric when finally installed can we properly plan for completion in the shortest possible time. It is the lack of appreciation of services interfaces and an inability to integrate their construction activity with those of the building works that can lead to an unforeseen hiatus, e.g. services works stop for fire spray on steelwork, or an area has to be vacated while the plant rooms are screeded. Situations such as these lead to claims, particularly on jobs where the programme shows M & E work as continuous. A little diligence in understanding BS interfaces prior to drawing up programmes can reap large benefits in productivity and site harmony.

Major interfaces for planning consideration are those where relatively minor value services work must be carried out in tight sequence with building work if delay to the latter is to be avoided; examples are enabling works such as the building in of puddle flanges, the provision of flue templates for casting bolts in bases, providing and/or fixing lightning protection connectors, tapes and earth rod pits.

The laying of underslab drainage in close coordination with the pouring of concrete bays is an example of a major interface occurring over a large area. Moving up the building, air intakes and discharges through the envelope and the building of riser shafts must be carefully considered so that building and services work can proceed with mutual efficiency. Externally the construction of ducts and trenches can play havoc with landscaping if not properly sequenced; see Table 5.8.

Table 5.8 Major interfaces (controlling activities) for building and services work

Puddle flanges, flue templates, earth rod pits	Enabling works (building progress)
Ingress/egress for utilities	Foundations, basements
Underslab drainage	Foundations, basements
Air intakes/discharges	Envelope
Internal penetration of structure, walls/floors	Release of work areas
Loading plant into plantrooms	Access; release of work areas
Ducts/trenches	Externals
Riser shafts	Release of work areas
Supports/fixings/secondary steel	Release of work areas

Minor interfaces need to be carefully searched out. If not given the right attention they can become a big problem. Particularly look for those situations where there are a number of repeated instances, e.g. location of socket outlets and switch plates in dry lined partitions. Taking this example further, electrical power run in conduit on the face of a blockwork wall studded out and plaster boarded may leave conduit (fixed to the blockwork) too far away from the socket outlet plate on the plasterboard surface. Who is to blame, the electrical design or the installing contractor? One will say 'Your working drawings should have picked up the difference between stud and conduit box depths', the other replying, 'You should have produced a co-ordination detail'. Followed by 'We weren't paid to do co-ordination.' Meanwhile the builder, although not looking at costs attributable to him (unless the stud depth was wrong?) is frustrated by the loss of progress that the rework will cause. In similar vein, look carefully for any specified requirement of services terminals needing to be coordinated with joint lines in tiled areas. This train of thought takes us into another specific area, of providing better facilities within buildings for disabled persons. The BS facilities disabled people need to use for access into and movement through buildings, and in toilet areas, are usually precisely dimensioned. If they are not, instructions should be sought from the DT.

Walls, floors and ceilings take us from the minor interfaces largely dictated by occupant usage of the services, into those related to structure and fabric module dimensions. Here we have seemingly impossible compatibility, at best limited rationalization. Consider Fig. 5.4(a) *integrated with (b) Services Outlet Modules*. The architect and structural engineer have little difficulty on reaching agreement on a structural grid of say 6 m to 9 m that will allow the architect to express his thoughts and achieve coherence and beauty in the external elevations. Within the structural grid further modular divisions of 1500 mm, 1200 mm or 1800 mm go naturally into smaller subdivisions of 500 mm or 600 mm. The architect has found reasonable freedom for the choice of wall, floor and ceiling elements. Very importantly the letting agent and client have their flexibility criteria for creating an acceptable variety of cellular offices. These too can be accommodated. Enter the villain. To create a suitable mechanically ventilated or air conditioned and artificially lit environment of optimum design the BS designer would wish to locate the outlets for those services in positions that rarely coincide with the architect's design model. The situation is made worse the greater the number of alternative internal partition layouts that the services designer has to cater for. Before we can make this better it may get worse. Legislation and/or the insurers may demand sprinkler protection. Much has, and can be done by the BS industry to accommodate modular layouts, but with the penalty of increased cost. The sprinkler design engineer cannot be so accommodating. To provide coverage according to the

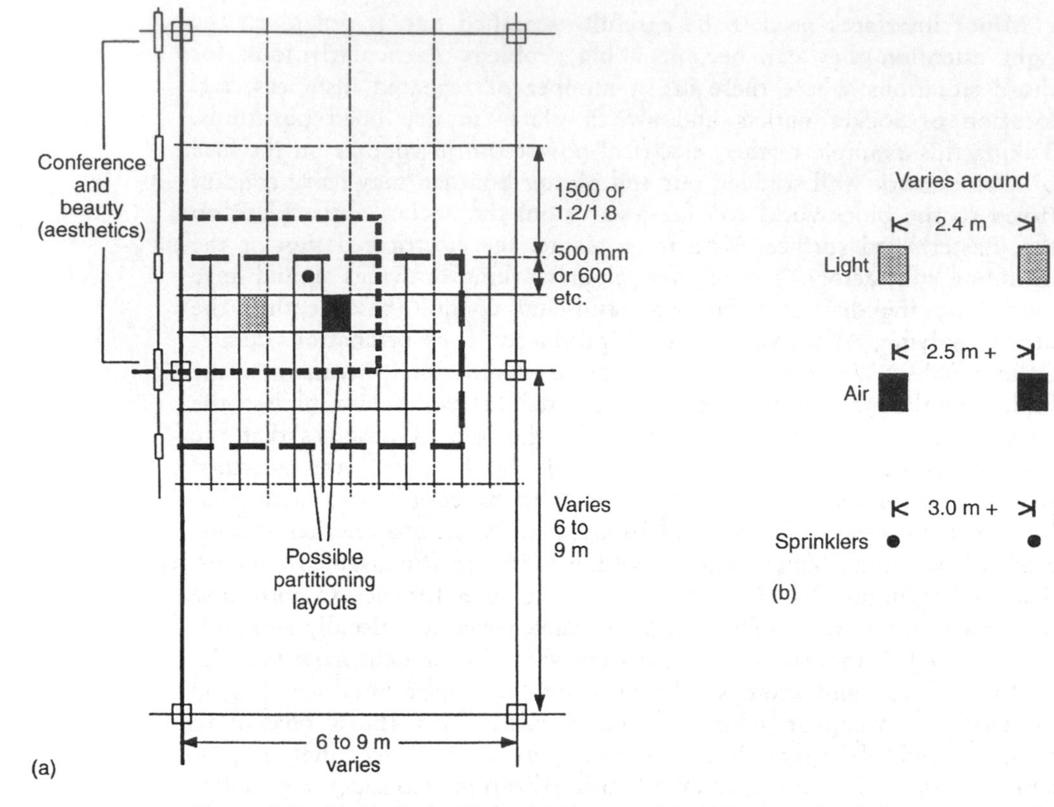

Figure 5.4 (a) Typical office planning modules; (b) typical office services outlet modules.

hazard rating for the risk category, partition layout changes can mean more heads in different positions. There is just about enough tolerance in sprinkler design for a head to be positioned in the centre of a 600 mm × 600 mm ceiling tile. To centre an outlet in a 1200 mm × 1200 mm tile may mean that in order to avoid being 'out of distance' it has to be offset to a degree that it will not provide adequate coverage to the area it is protecting. Change sprinklers for heat or smoke detectors and the situation is similar. Passive detection and protection systems take priority in the discharge of their function over accommodating building modularity.

Due to its brittleness the building module is king:

- It limits options.
- It requires rigid coordination.
- It supports abuse through bad management/coordination.
- It repays careful study in programming construction.

Table 5.9 Example of special interfaces – security services

Card access – door contacts and wiring
Door phone – door contacts and wiring
Wall tremblers } intruder alarms
IR sensors } intruder alarms
Volume change sensors
Location of CCTV cameras
Traffic barriers
External lighting

Special interfaces are the wild cards of which security systems are a good example. Look at Table 5.9. Not all, of course, are provided in every building. The pattern of provision through the building may sometimes be detected from studying the drawings and specification, e.g. for card access along circulation routes into different departments. Generally, though, it is security services relationship to the geography of the site and the building layout that must be studied in this example of special interfaces. The door phone that is provided at entry to a block of social housing accommodation may be the one piece of 'high tech' in what is otherwise domestic-scale building services. Tremblers embedded in the external walls will need to find a route for the terminal wiring. The location of CCTV cameras may require a dedicated duct or trench for wiring through the hard or soft landscaped areas. Similar interfaces occur with external lighting where boundaries between what is particular for security or safety can become blurred. Taking both together they fall naturally into two groups. Lighting mounted off the building creates interfaces in its BWIC and access for fixing the terminals which may need to be brought forward for installation before scaffolding is struck. If not considered in depth, one can be involved in the expense of mobile access towers or waiting until the facade maintenance equipment is operational on very tall buildings. The second group are those positioned externally along the site's traffic and pedestrian routes, for car parks and flood lighting. It is a pity not only for the time and cost lost, but for the visually unattractive, albeit temporarily until weathered, look of remedial work to hard and soft landscaping.

5.3.4 ACTIVITIES LISTS

Armed with the knowledge of what services are to be installed, where they are located, and a clear view of their interface with the building works, the activities for BS construction are easier to create. Unfortunately this cannot always be left to the BS firm. A lack of ability to provide adequate planning has a number of route causes: builders may be uninterested in doing the job properly on traditional contracts for reasons stated earlier; management contractors may hijack all planning activities telling the BS contractor exactly where to be and what to be doing during a particular week. In between there is a fair amount of mediocrity. Those BS firms that plan well are regrettably few. To the contractual route reasoning one can add the effect of the position of the economic pendulum. Recession reduces management resources to frayed shoestrings, a position from which, return to the security of a belt, and the golden era of braces as well, is perhaps unimaginable. The lack of skill and attention applied to the planning of building services is the area of greatest managerial shortfall. By reversing the point it becomes the area for greatest improvement in which the builder and BS contractors need to come together if reduction in building costs is to be achieved with increased profitability for both. Are you not sick of being told that the Japanese build quicker and better than we do? They don't! They spend more time planning and programming, starting in design and continuing in depth down to the toolbox meeting.

The riding of this particular hobby horse has not occurred here in 'Activities' by accident. The defining of activities must be done in a depth that is appropriate to each job. Some projects are simple with domestic-scale building services. Others are far more complex; the activity lists will be stepping stones to sequences with other trades, access to plant areas, high and low level sequences and resource cycles, labour gangs and myriad interdependencies.

Appendix I shows a range of BS activities. It is not exhaustive, and its application will vary according to the building function, form and location. It has been created in a matrix of first fix, second fix, final connections and terminals under the following headings:

- externals
- on building face
- internals (for water tightness)
- plant areas
- mechanical services
- electrical services
- controls
- range of mainly second fix and final activities (all services)
- telecoms and data
- generators and UPS
- special ancillary electrical services (second fix, final connections and test)
- lifts.

Where the list does not provide sufficient detail for a specific project it is hoped that it will be a framework for developing adequate listings.

Some activities occur in both second fix, and final connections and terminals columns, for example 'position and fix sanitary fittings'. In work on domestic housing it is possible that because the house or flat can be locked after activity, for the sanitary fittings, seats, plugs, chains and tap tops to be carried out all at the same time. Where access cannot be so controlled the fitting of seats, plugs, chains, etc., will be last moment activities to be closely followed by 'final clean'. Another example of variable activity placement is in earthing and bonding. Most usually carried out progressively this may be subject to 'follow up' work teams, or where the installation remains exposed, left as a 'finals' activity.

There are situations where we need to take our knowledge of the content of first fix, second fix and finals a little further. Once again domestic type work gives us an insight. Study Fig. 5.5, in which the insulating effect of quilt precludes the use of twin and earth wiring and necessitates conduit. Elsewhere on the same job partitions occurred without the fibreglass quilt. The differing construction affects the content of the electrical activities in the following way:

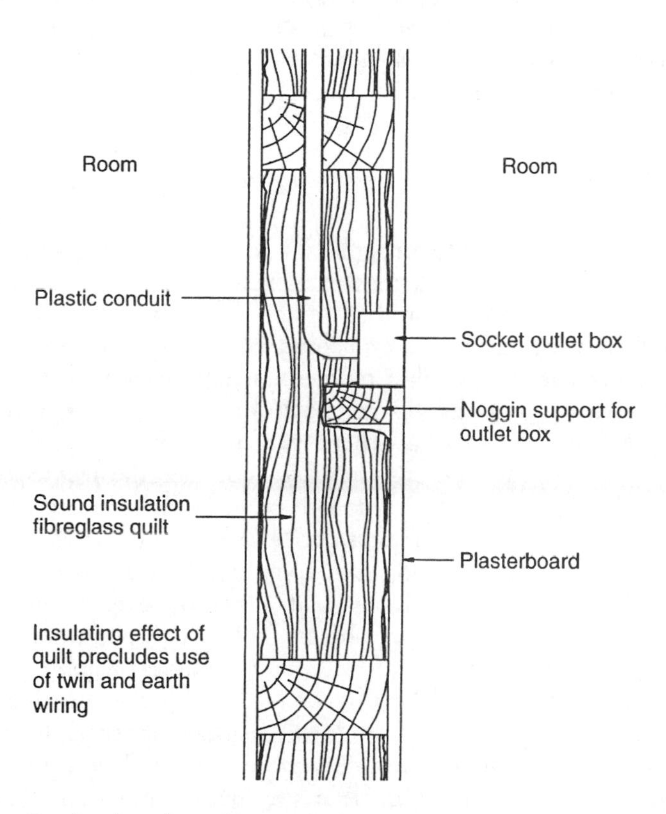

Figure 5.5 Dry lined stud partition – section.

Construction	*1st fix*	*2nd fix*	*Finals*
Uninsulated partition	Twin and earth	Terminals	–
Insulated partition	Conduit and draw wire	Wire up	fix terminals

Accepting that the conduit box is part of the first fix for both constructions the insulated partition imposes a three activity sequence. Without understanding this the site manager can be led into misunderstanding the extent of progress on the electrical installation. The twin and earth would be run before the completion of the plaster boarding whereas the wiring out of the conduit is most likely to be carried out after boarding out.

First fix activities are sometimes referred to as 'carcassing'. Whatever title the work masquerades under, knowledge of what is actually included is important, sometimes critical. Some contractors define first fix, second fix and third fix (finals) with inspection and witnessing hold points. N.G. Bailey has kindly provided the activity category examples in Fig. 5.6 (a) (b). BS contractors should provide similar listings with their method statement. If contractors are reluctant to provide this information they can be questioned as to, 'How did you assess the safety hazards without defining the content of the work?' Pursuing the safety line usually elicits an adequate response.

5.3.5 SEQUENCE

It has taken a long time to progress to this point, the point at which most planners dive in straight away. Their philosophy of having the rainwater disposal system and sanitary plumbing roof penetration pipework follow upwards closely behind construction in order to 'seal' the roof at the earliest possible moment, makes sound sense. Coupling this with insisting that the BS contractor starts with the lowest level plant installations as soon as there are a couple of levels of floor construction overhead, weakens the case. Whether the frame is in steel or reinforced concrete (RC) and construction work is cleared from two or three levels above, rarely makes for 'weathertight' plant areas at the lower levels. Riser shafts and numerous small floor penetrations for services will require 'brick edging' and a board cover to improve the situation. Without plant rooms laid to falls with operational drainage gullies, or permanent sump pits from which temporary sump pumps can drain off the water, we will be forever faced with the 'floating scaffold board' access route to inspect shrouded, desiccant laden chillers, rusting boilers and sweaty switchboards. To avoid confrontation and disputes over protection responsibilities, the site manager is advised to get involved with the BS contractor so that both parties are aware of what is expected from the other and define acceptable conditions for working areas. It is good practice to get the BS contractor into plantroom areas which he can 'own' and control access to. A number of specialists will also require dedicated plant areas handed over to them as soon as possible, e.g. diesel generator and UPS rooms, lift shafts and machine rooms. It is well worth the site manager finding out when the specialists want their areas rather than telling them when they get them, only to find out to near to the end of the job that work will not be complete because 'You gave us our plant rooms and work areas too late' – 'But you didn't tell me' – 'But you didn't ask me' – result conflict, cause lack of communication.

Outside of the plant room areas the distribution systems of building services have their own installation pecking order. A typical priority of position and level would be:

- rainwater outlets and pipes to drain the external run-off areas;
- plumbing stacks go next, not only because they too may penetrate the roof and will need flashing in, but any offsets have falls that take priority over adjacent services which must set round them;
- heavy cabling with limited bending radii needs space for pulling up ladder rack or tray;
- ducting usually comes next but depending on the system, i.e. type of air conditioning – VAV terminals, FCUs and VRV cassettes may precede ducting. Ducting takes precedence over pipework, not only because of its size, but its limited ability to offset around other services without possibly creating noise and certainly increasing the system frictional resistance above that which the fans have been sized to overcome;
- pipework, particularly that associated with water heating, has a greater capability below say 100 mm diameter to offset around the preceding services listed above. Although this too has difficulties of noise generation, increased resistance and additional venting and draining to be considered, it is nevertheless the service that usually has to give way. Sprinkler pipework running horizontally will follow in false ceilings below the other services and probably run just above ceiling membrane;
- the horizontal method of distributing electrical services affects when they are installed, e.g. a conduit distribution running on the soffitt of a slab may go in before the HVAC ducting. This could also apply to tray and trunking work above the false ceiling. Running in a floor void the electrical distribution, probably now embracing data and telecoms will take priority. In this instance, pipework running out to serve some form of building perimeter heating would do the offsetting.

These then are the generalities, the framework around which the site manager and planning support will consider the needs of their particular

Bailey — **MECHANICAL VISUAL INSPECTION RECORD**

CLIENT WC (UK) Limited AREA ____ CONTRACT NAME Kingspool Development CONTRACT Nº G6/66024

FIRST FIX		SECOND FIX		THIRD FIX	
ITEM	INSPECTED	ITEM	INSPECTED	ITEM	INSPECTED
Support Brackets adequate Fixings adequate Pipework:- - Size correct - Class - Bracket Spacing - Routing - Welding - Pressure Testing - Expansion Provision - Sleeves - Venting Provision Ductwork:- - Continuous - Mechanically sound - Fire Dampers installed - Access installed - VCD's installed - Anti Vibration isolation Snag Sheets raised (NGB 084) Snag Sheets cleared		Pipework:- - Size correct - Class - Bracket Spacing - Pressure Testing - Expansion/Venting Provision Ductwork:- - Continuous & Mech sound - Fire Dampers installed - Access installed - VCD's installed Heat Emitters:- - Radiators correctly installed - Fan Coils correctly installed - Trench Htg correctly installed Waterside Plant:- - Located correctly Airside Plant:- - Located correctly Control Panels located Snag Sheets raised (NGB 084) Snag Sheets cleared		Ductwork:- - Flex/Grilles fitted Pipework:- - Soundness check Heat Emitters:- - Condition check Rads/Fan Coils - TRV Head fitted - Grilles fitted on Fan Coils Waterside Plant:- - Accessories fitted - Flexible Conns fitted - Control Items fitted Airside Plant:- - Accessories fitted - Flexible Conns fitted Check general conditions of Plant Items Rotation check of Fans/Pumps Insulation installed and ID Labels/Notices/ Markings/Diagrams/Charts/Instructions Snag Sheets raised (NGB 084) Snag Sheets cleared	

INSPECTED BY:	WITNESSED BY:	INSPECTED BY:	WITNESSED BY:	INSPECTED BY:	WITNESSED BY:
Name Print ____	Name Print ____	Name Print ____	Name Print ____	Name Print ____	Name Print ____
Name Sign ____	Name Sign ____	Name Sign ____	Name Sign ____	Name Sign ____	Name Sign ____
Date ____	Date ____	Date ____	Date ____	Date ____	Date ____

NGB 080 MAR 94

(a)

Figure 5.6(a) Activity category lists on mechanical visual inspection record from N.G. Bailey & Co Ltd.

Bailey

ELECTRICAL VISUAL INSPECTION RECORD

CLIENT WC (UK) Limited AREA ______ CONTRACT NAME Kingspool Development CONTRACT Nº G6/66024

FIRST FIX		SECOND FIX		THIRD FIX	
ITEM	INSPECTED	ITEM	INSPECTED	ITEM	INSPECTED
Support brackets adequate Fixings adequate Conduit/Trunking:- - Class - Couplings/bushes - Saddles/clips - Bends/sets - Routing - Burrs removed - Earth continuity Traywork:- - Continuous - Mechanically sound Cable tiles/tape installed Ducting completed Corrosion protection Installation methods satisfactory Snag Sheets raised (NGB 084) Snag Sheets cleared		Cable:- Type correct Grouping Bending radius Cleating/clips Size correct Routing Protection Earthing/Bonding:- - Main earth termination - Main earth bar - Main equipotential bonds - Supplementary bonds Steelwork continuity tested Distribution Equipment:- - Securely fixed - Positioned correctly - Size/rating adequate - Installation completed Snag Sheets raised (NGB 084) Snag Sheets cleared		Cable:- - Connection of conductors - Identification of conductors - Polarity - Heat sleeved Insulation of live parts Barriers/enclosures/obstacles Protective conductors installed Earthing/bonding conductors terminated Isolation/switching adequate Protective device ratings adequate Labels/notices/markings Diagrams/charts/instructions Lids/covers fitted IP ratings adequate Apparatus rating adequate Fire barriers fitted Accessories fitted Snag Sheets raised (NGB 084) Snag Sheets cleared	
INSPECTED BY: Name Print ______ Name Sign ______ Date ______	WITNESSED BY: Name Print ______ Name Sign ______ Date ______	INSPECTED BY: Name Print ______ Name Sign ______ Date ______	WITNESSED BY: Name Print ______ Name Sign ______ Date ______	INSPECTED BY: Name Print ______ Name Sign ______ Date ______	WITNESSED BY: Name Print ______ Name Sign ______ Date ______

NGB 039 MAR 94

(b)

Figure 5.6(b) Activity category lists on electrical visual inspection record from N.G. Bailey & Co Ltd.

project which may change the running order. Projects such as hospitals, laboratories, silicon chip wafer fabrication facilities with dominating street services management principles will create their own order.

5.3.6 HIGH LEVEL AND LOW LEVEL SEQUENCES

Previously we touched on the sequencing of services above false ceilings and below false floors. Site managers who have achieved success doing high level first and low level second should still be wary of the next project where the right way to do it could be to reverse the order of the last success. Proponents for the 'Always do the floor work first' argue that the floor can be protected; they extend the argument by pointing out that ceiling work takes a lot longer to finish because the ceiling is always being re-entered for testing and commissioning the HVAC. This is true; for the new building fully fitted out, services below the floor could be the way to go. For refurbishment, or fit out of a 'shell and core' office, high level first is shown in Fig. 5.7. This sequence was chosen because there were fewer high level activities which could be carried out more quickly and the VAV risers to the entry of the office floor had been properly commissioned. If there were going to be any variations it was anticipated that these would occur to the partition layout and floor boxes serving electronic workstations which would all be low level activities. So it proved in practice.

5.3.7 RESOURCES

With the knowledge gained from the thorough examination and understanding of the BS and having them properly integrated with the building trades the site manager should be comfortable in proceeding with his planning support to discuss with the BS contractor the proper resourcing of their work. The single line bar chart for M & E or a few more lines for first fix, second fix and commissioning is a brushstroke approach over which no intelligent discussion can take place. Without proper planning the site's health and welfare facilities may be put at risk when a visiting H & S inspector correlates labour returns with site facilities. You have no answer; you don't want the BS contractor to reduce his labour which you have just persuaded him to increase because you thought (but could not prove from the sketchy programme information) that he was well behind programme. With good programming the site manager has the best of all monitoring tools and is enabled to pinpoint any slippage. Recovery is possible in a controlled way and expenditure quantified; even it is someone else's

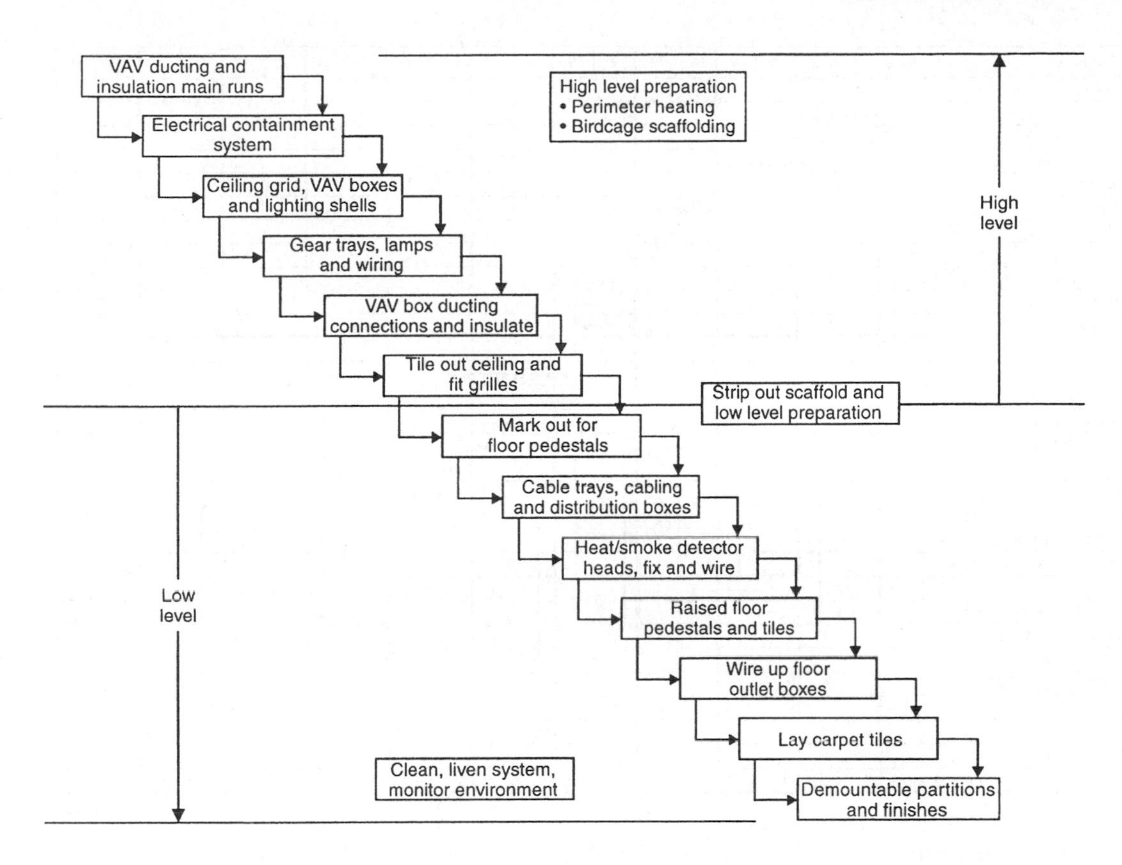

Figure 5.7 High level/low level fit out sequence for typical office floor.

loss it is under better control. Naturally the site manager will want to see a smooth build up and run down of labour resources; but be wary of refurbishment and fit out work for they can cause peaks and troughs about which very little can be done. See Fig. 5.8 (derived from Fig. 5.7) on which the labour histogram is superimposed. This 'city skyline' labour profile has no scope for 'resource smoothing' on an individual floor basis but, by looking at all the floors in five blocks, it was possible to achieve steady build up to a plateau workforce and sensible decay.

Controlled resourcing is important in that it must be matched by adequate plant, equipment, toolboxes, e.g. adequate numbers of temporary flexible connections, pressure pumps and gauges. How much better this is than thumping the table, being given labour whose productivity is negligible as they rush from floor to floor, sharing tools, bending machines and pipe cutters.

The message is plan, plan and plan again.

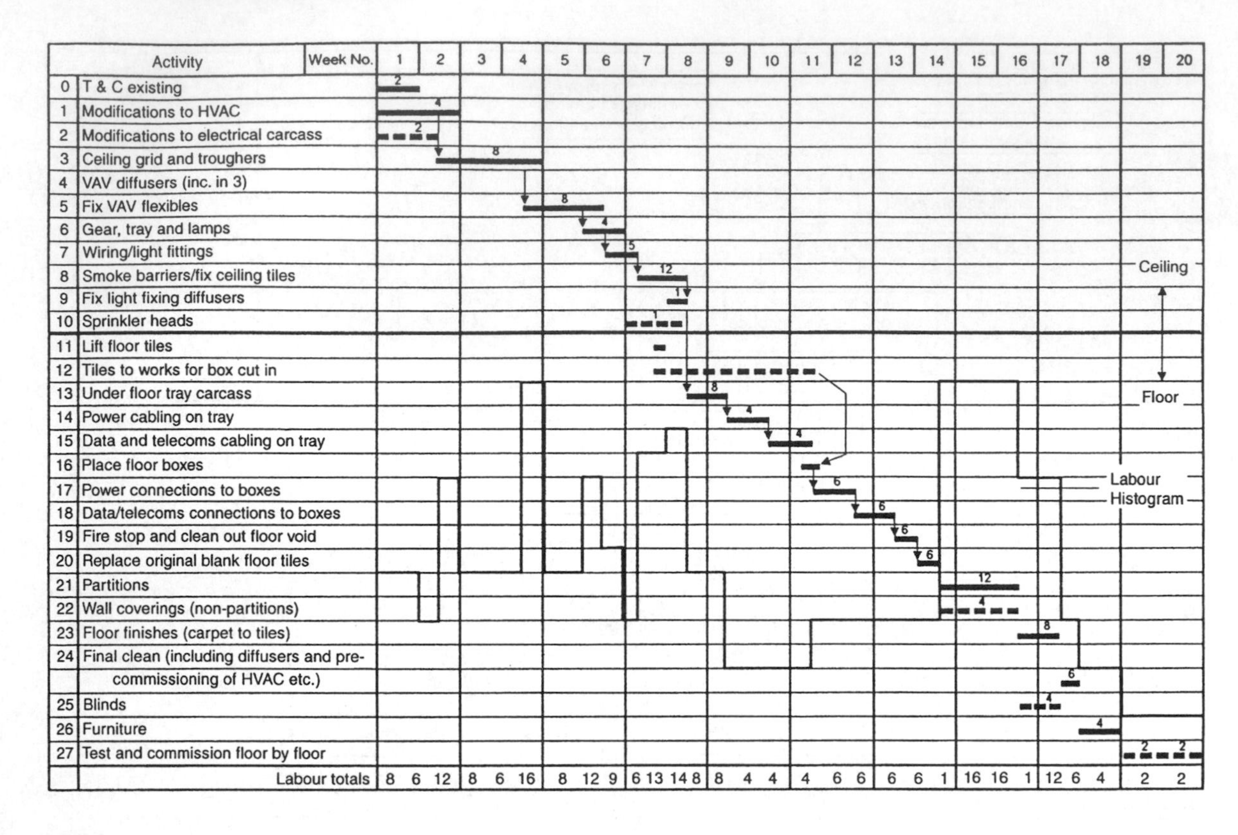

Figure 5.8 Typical programme for fit out of 1000 m^2 office floor.

Schedules 6

6.1 Schedules, what schedules?

This subject may strike a discordant note and appear to be tangential to the logical flow of managing a services contractor from pre-award to handover and beyond. Nevertheless a careful study of the information contained within schedules will reward the site manager with much useful knowledge. For it is the way in which schedules are looked at that creates knowledge from what are often apparently strings of information. So we will apply some orderly thought to what a site manager might learn from:

- schedules at tender stage;
- schedules created pre-construction; and
- schedules created during construction.

Schedules will be given further treatment according to their place in the management of commissioning, defects and handover; see Chapters 9, 10 and 11.

6.2 Schedules existing at tender stage

The DT's schedules will have been studied in a particular way in order to win the job. Schedules in bills of quantities, in specifications and on drawings listing plant, equipment and materials will have been estimated by the BS contractor, who will have studied the schedules and quantified their requirements. Some of that knowledge should by now have been imparted to the building site manager and planner in dialogue with the BS contractor during programming.

In section 5.2 the site manager and planner were exhorted to identify services, and look for interfaces with structure and fabric. They were told to look for major, minor, special interfaces and those with wall, floor and ceiling finishes. To do this diligently they will have studied the DT's documentation issued both at tender enquiry, and 'for construction'. In looking at this material again not for the purpose of tendering, but now for constructing the work, it is possible they will discover illuminating detail that takes on a different significance. Look at Tables 6.1–6.3 for the information they provide and the questions raised, now to be resolved with the BS or specialist contractor.

Table 6.1 provided information to the effect that security on this job was extensive, as we were looking at page 5 of 6. It raised a question as

Table 6.1 Petrol filling station camera and monitor schedules (sheet 5 of 6)

Item	Location	Type	Comment
CCTV camera schedule			
1	Canopy (external)	Fixed lens (Panasonic WV-BL90 or alternatively JVC TK-S130E)	CCD camera mounted on underside of canopy, weatherproof housing with external lens
2	Canopy (external)	Fixed lens (Panasonic WV-BL90 or alternatively JVC TK-S130E)	CCD camera mounted on underside of canopy, weatherproof housing with external lens
3	Forecourt	Fixed Lens (Panasonic WV-BL90 or alternatively JVC TK-S130E)	CCD camera mounted on 4 metre column weatherproof housing with external lens
4	Kiosk (internal)	Fixed lens	CCD camera wall mounted
5	Kiosk (internal)	Fixed lens	CCD camera wall mounted
Monitor schedule			
1	Kiosk (internal)	1 × 9" (Panasonic WV-BM90 or alternatively JVC TM-9060E)	Mounted on a swivel plate on desk top counter
2	Kiosk (internal)	1 × 9" (Panasonic WV-BM90 or alternatively JVC TM-9060E)	Mounted on a swivel plate on desk top counter

Table 6.2 Schedule of louvres

Plantroom	L × H
L1	4400 × 3100
L2	4400 × 3100
L3	4400 × 3100
L5	4000 × 450
L6	4000 × 525
L7	4000 × 675
L8	4000 × 900
L9	3200 × 1125

Table 6.3 Schedule of specified emergency luminaires

Ref	LAMP	DESCRIPTION	MANUFACTURER	CAT REF
B1	F	3 hour conversion pack for use with type A1 luminaires	Menvier Ltd	–
B2	F	Dichroic conversion pack for use with type A2/A3 luminaires	Bradley Lomas Electrolok Ltd	DC50
B3	F	Self-contained non-maintained emergency exit sign complete with 8 W fluorescent lamp with Exit panel	Bradley Lomas Electrolok Ltd	AEU/ICEL/5
B4	F	Self-contained non-maintained emergency luminaire complete with 8 W fluorescent lamp, opal poly-carbonate diffuser protected to IP65	Bradley Lomas Electrolok Ltd	EU83

Notes: 1. Where indicated on the drawings, general lighting luminaires shall be factory converted to provide 3 hour emergency usage. Conversion packs shall be mounted remote from the luminaires is sealed enclosures within ceiling voids and manufactured by Menvier Ltd, their Convertalite range.

2. Low voltage luminaires shall be converted for emergency usage where shown on the drawings by remote conversion packs, as manufactured by Bradley Lomas Electrolok Ltd, model No. DC50 mounted in separate enclosures within ceiling voids.

to who should provide the 4 m column for item 3, forecourt, and for items 1 and 2; it also raised the question as to how the cameras were mounted on the undersides of the canopies.

This schedule in Table 6.2 could be misleading in its simplicity. There are nine louvres and they are large. This schedule was in the BS specification for that contractor to supply. The questions raised and to be discussed with the BS contractor, if not answered in the descriptive part of the specification, concern fixings: how and by whom? Is the opening steelwork trimmed? And lined with a hardwood subframe? Are the louvres to be delivered in sections? Who is to hoist them into position? Finally, before leaving this figure it is worth asking the architect if the specified finish is correct. It is not unknown for the finish of items such as these, located as they are in the external elevation of the building, but specified by the services consultant, to be different from those the architect/client thought they were getting. For louvres to have to return to works leaves the site manager without a credible weathertight building. This can lead to conflict over the costs for instructions for temporary weathering.

The first thing to note in Table 6.3 is that these units are special. According to note 1 general lighting luminaires are to be converted in the factory. According to note 2 B2 is to be supplied with 'remote conversion packs'. It would appear that the pack itself is standard as it has a manufacturer's identity number, DC50. But are the packs mounted in separate enclosures specials? Plenty of scope here for these to be missed. And, most importantly for the builder the pack in its enclosure is mounted in the ceiling void 'separate' from the luminaire. It is most likely that the luminaire location is clearly identified; but what of this separate enclosure? Undirected, the electrician will most probably use his initiative and fix it how and where he can in a position that is most suitable for him, without consideration of accessibility for maintenance or repair. So what appeared to be an innocent looking schedule becomes one of considerable importance to workface coordination, particularly if the emergency luminaires are in different types of rooms. In that case each one is a different coordination problem as to where these separate 'enclosures' are to be positioned in the ceiling void.

6.3 Schedules created pre-construction

Well in advance of start on site the site manager/planner should have called for the BS contractor to programme the production of information for which the source material would be schedules of the same name as the agenda items – see Table 5.1. Remember all of the schedules covering the delivery of information and subsequent output is required for each BS contractor, some of whom may be specialists whose start on site is later than the general BS contractor(s). It is easy to have scheduled all of the pre-construction activities for the bulk of the services work and for that to be in full swing, but forget later specialist contractors who must be taken through the same pre-construction process.

There is at least one other important pre-construction list to be prepared and that is the hoisting schedule. Here we must accept that the site manager is competent, and on the larger projects, ably supported by a construction manager also well versed in matters pertaining to the movement of plant, equipment, people and materials up, into and through buildings under construction. Nevertheless, many site managers on management contracts have been taken unawares by the BS contractor who arrives with a low loader and hired crane accompanying the delivery of e.g. the boilers, for 'dropping into the basement', or for chillers to be placed on the roof. Do not be trapped by the contractual arrangement that places hoisting responsibilities on each works or trade contractor. If you want your site to be smooth running then whatever the contract form, as site manager you must get involved in coordinating plant, equipment and material deliveries that involve hoisting. For all contracts the site manager should call up schedules from the BS contractor that give each item details of volume, packed size and weight. The trigger for requesting this information should be delivery of information schedules listing offsite manufacture, and procurement. It is at the planning stage that any need, scope and attributable costs for having plant and equipment delivered in smaller/lighter loads should be finalized. Of course, requirements should all have been made known to the BS contractor at tender stage, but sometimes the site realities negate tender theories. Hoisting is a subject that must be revisited post-tender and pre-construction for each services contract.

6.4 Schedules created during construction

The importance of scheduling cannot be overemphasized and those schedules of construction covering:

- the status of installation
- the status of construction testing
- instructions (including variations) and
- defects

will be positioned in practice in the following stages of controlling the works.

Supervision and inspection 7

7.1 Control of work

Programming has given direction to our strategy for construction and we can approach the workface confidently. Easy to say, but do we have a right to feel confident? How well are we managing? Let us look at our definition of management – planning, organization, coordination and control, and examine where we are:

- Planning: this has been done through the proper programming of the delivery of information and construction activities.
- Organization: information has been prepared, e.g. quality, safety and environmental plans, methods and safe working practices set down and approved, samples submitted, offsite production is under way and the workforce mobilized.
- Coordination has been demonstrated in the programme by setting out an activities sequence, and determining the content of services first and second fix in relation to each other and preceding and following trades.
- Control: yes, we are right to be confident; so much has been done to assist on-site production that control should be easy.

Regrettably, control of the work is not conducted properly due to misunderstanding of what is involved. Before we go any further it is important to understand more definitions.

- Supervision: instructing before and during an activity;
- Inspection: examination during and/or after an activity;
- Quality Control: compliance with specifications, codes of practice, etc., at the workface (where specified requirements must ultimately be achieved), verified by test and inspection.

Most builders have, or should have, statements in their QMS standing instructions of how work should be controlled along the following lines:

> All work carried out by Contractors shall be controlled by their own management systems which shall be approved by the Company. Contractors' Quality Plans shall include for Method Statements defining how the work is to be carried out and the

> appropriate Test and Inspection Plans clearly identifying critical stages of the work by *hold points* beyond which work shall not be allowed to proceed until the preceding stage has been checked and assurance provided that the work conforms to the requirements.

With clear simple definitions for supervision and inspection coupled with a firm statement for controlling the work we can proceed to consider the subject in greater detail.

In recognition of the impact that the CDM regulations have, from this point onwards the builder will be referred to as the PC and all other subcontractors and specialists, irrespective of how they may be termed under the contractual arrangement of their employment on a project, will be referred to as contractors, e.g. BS contractor.

7.2 Division of responsibility (for supervision and inspection)

7.2.1 GENERAL

In his regular column in *Building*, 11 February 1994, Colin Harding, head of Bournemouth based contractor George and Harding, under an article entitled, 'Time for architects to relinquish their leadership' [1] described the origins of supervision in the 1923 Form of Contract that: 'Divided our industry into two distinct halves: the "supervised" – contractors, sub contractors and specialist suppliers; and the "supervisors" – architects, engineers and quantity surveyors.'

In the same article Harding goes on to describe the changes in the abilities in the two halves of the industry that have occurred

> The old guard supervisors (architects, quantity surveyors and specialist consultants) have been getting gradually less efficient and confident in their project management role. At the same time the old 'supervised' builders have increased their management skills, raised their standards of education and training (at the peak, builders were recruiting 900 graduates a year), and have become more professional and conscious of their clients' needs. We now have chartered building companies with Codes of Conduct as demanding as any professional institution.

Writing earlier in *Building Services*, February 1987 in an article entitled 'The building site minefield', [2] Jack Torrance, a past president of CIBSE states: 'It seems to me that the whole question of supervision is so fraught with uncertainties and misunderstandings'. In his article Torrance describes the difficulty of marking a safe route through the minefield of supervision due to misunderstandings in needs and expectations and duties placed or not placed by the client and his advisers. With Harding's view of the builder's competency and Torrance's description of the pitfalls we will proceed with some basic thoughts on the responsibility for supervision and inspection by the parties involved in a project. Difficult though this may be, the site manager would do well to acquire as clear an understanding as possible of the defined supervisory and inspection roles that each party has been signed up to perform.

7.2.2 THE CLIENT

The client who is a regular procurer of buildings should be well able to set out the marker flags through the supervision and inspection minefield with some safety. The marker flags will be positioned in conjunction with his own or externally appointed project manager and the design team leader. According to the contractual form the employer's representative, architect or contract administrator will be assigned responsibility for issuing instructions. It is well known that those projects enjoying the regular involvement of the client's 'in house' representative proceed with the least difficulty.

For the site manager who is constructing a job for the 'once in twenty years', client, there are warning signs. For the inexperienced lonely client given what sounds like good advice from quantity surveyors and architects lacking project management skills, come the origins of poorly assigned supervision and inspection roles. Beware, it is these clients who, through no fault of their own, when engaging an architect to 'supervise and inspect the works' will expect his appointee to be on site full time and look at everything.

7.2.3 THE DESIGN TEAM

Working for the informed client or appointed by an experienced PM who can carry the case for making other appointments for him, an architect or DT coordinator may have little difficulty in getting resident engineers and building clerk-of-works on board. Fine for the architect and structural engineer. As Torrance explains, the M & E consultant is less fortunate when he seeks to appoint a services clerk-of-works:

> The Architect who should support the proposal rarely shows any enthusiasm for it to the Client, and the Client anyway believes he already pays for supervision in the fees to the Consulting Engineer. So, why pay more?

Members of the DT can usually only instruct by requesting the architect or contract administrator (often one and the same) to issue an instruction. Sometimes, but unusually so, BS consultants are empowered under the contract to issue an engineer's instruction. The scope of an engineer's instructions (EI) is limited to matters of building services workmanship. The site manager should be wary of any EI that involves trades other than building services. Unless the site manager is certain that the architect or structural engineer is in agreement with the EI, he may find himself entangled in DT differences at a cost to his project that will be difficult to recover.

7.2.4 CLERK-OF-WORKS

Although they come in all shapes and sizes there are only two types of clerks-of-works, for buildings or M & E. Irrespective of skills in these disciplines clerks-of-works fall generally into two categories: helpful or unhelpful. It is not usual for clerks-of-works to be empowered to give instructions and yet in their inspecting role the lists of defects they make are implied instructions – 'the following do not comply with the specification' – and are therefore instructions for corrective action.

The helpful BS clerk-of-works may try the 'word in the ear' approach and allow a little time before judging that the BS contractor is trying to avoid compliance, only then recording the aberrations. This clerk-of-works is also likely to contribute to the solving of technical problems, turning a blind eye when, for example, brackets on cable, pipe and duct distribution systems are not exactly to specification, but still comply with that indefinable good engineering practice. Nor will this clerk-of-works run to his employer, the M & E designer, to get a 'bad news fax' issued. No, this clerk-of-works is responsible, proactive and his support, which is paid for by someone else, is well worth cultivating by the site manager. The other type of BS clerk-of-works is the opposite. Pedantically thorough at best, clerks-of-works of this type can reduce a services contractor's productivity through their constant questioning about what is going to happen in the future:

- 'When are you pulling cables into the switch room?'
- 'How are you protecting the boilers?'
- 'Are you bringing the lift car in in one piece?' or the worrisome 'I've never used black enamel conduit in a lift motor room before, I'm trying to get the engineer to change it to galvanized.'

Add to this the relay of tittle-tattle back to the engineer's office about matters that you, the site manager, are dealing with concerning BS resources, time keeping or say the use of personal protective equipment and we have a clerk-of-works making a negative contribution to the site. At the risk of making a bad situation worse the site manager may need to have the clerk-of-work's terms of reference defined and limited as to what he can and cannot discuss with the BS contractor.

On those sites where the building clerk-of-works also has to cover the

BS works, we have a situation where responsibilities are attempted to be discharged by inappropriately qualified appointees. Even where the clerk-of-works is of a helpful nature, as he seeks to become informed, he will slow down the organization through the constant stream of questions. As site staff seek to cooperate with the clerk-of-works with the helpful attitude, they will most probably spend more time with him than with the person of the unhelpful attitude whose lack of services knowledge and pleas for enlightenment will be shunned.

7.2.5 PRINCIPAL CONTRACTOR

Whatever contractual arrangement is in place traditional, management, construction management or D & B the PC gives the instructions. Instructions are the instruments of supervision and simply put, take two forms. First they are instructions received by the PC, and secondly those he wishes and has a right to give under the contract. The instructions a PC receives come from two routes, one of which is the client/DT and the other from what we will call here the external inspectorate. The former will arrive in the standard procedural format of an architect's or engineer's instruction (AI or EI) respectively. From the external inspectorate, including local and supply authorities and possibly other agencies such as insurers, instructions may come in letter or official notification form. With respect to both types it is for the PC's site manager to decide what instructions are to be passed on to the contractors and in what form.

7.2.6 THE BS CONTRACTOR

The BS contractor will receive his instructions from the PC and also directly from the external inspectorate. It is possible that some members of this group that we have called the external inspectorate, e.g. gas, water and electrical supply companies may be in contract with the PC or BS contractor. Where it is the latter the BS contractor should pass to the PC any instruction received. Instruction may be implied through an inspection report, e.g. 'The plastic water main has been backfilled with gravel containing sharp stones. It should be backfilled as specified with pea shingle.' Here is a case where the water company has commented to the BS contractor on BWIC carried out by a PC.

7.2.7 EXTERNAL INSPECTORATE

Although we have already instanced some of the constituents of this group it needs to be defined more clearly. These are the inspectors and officers of the local authorities and regional utility companies and a few more organizations besides. Titles will vary for similar functions and therefore the listing below is not exhaustive.

- The inspectors: Building inspector, Sanitary inspector, H & S inspector;
- The officers: building control officer, fire control officer, environmental health officer.
- The regional utility companies: regional electrical company, regional water company and regional gas company.

The purveyors of telecom and data services by cable and radio consider themselves national although this is not always geographically correct.

To all of these who will come during and at the end of the job we can add client and tenant insurers who will show a particular interest in the verification of fire systems, lift installations and pressure vessels, e.g. boilers and cylinders.

The site manager should set up a project procedure that ensures that the PC is informed of any communication received by the BS contractor or his subtraders from the external inspectorate.

7.3 Supervision

7.3.1 GENERAL

Supervision is one of the two instruments for controlling the works, inspection being the other. Supervision gives direction to work. It may precede work, e.g. 'Start installing radiators on the 2nd floor next Monday' or arise from inspecting an activity during or after its construction, 'Your fitter on the 2nd Level south face dropped a radiator when trying to position it. Make sure he has a mate working with him' – notice of unsafe practice and an instruction during an activity. Or, 'Some of the connections to the radiators on Level 2 are like a dog's hind leg and need correcting' – an example of an instruction as a result of an inspection after the completed activity. What the site manager wants to see from his BS contractor is for supervision to be right: the Right

- work
- place
- time
- materials
- resources
- method.

Others will also be showing an interest in this under their responsibilities for supervision and along with the site manager will issue instructions if all is not well.

7.3.2 INSTRUCTIONS, THE CLIENT AND DESIGN TEAM INSTRUMENTS OF SUPERVISION

We have seen how the parties involved in a project issue instructions down the routes of communication afforded to them by their responsibilities under terms of engagement and contracts. By far the most common route down which instructions are issued is that from the client and DT. Perhaps 90% or more of all instructions received by the PC flow down this route. Starting with the order, drawings, specification and schedules to be used in construction, and requesting information comprising regular, incident and status reports. This route is also the conduit for instructing variations to the contract works requesting more or less, but always different work.

7.3.3 INSTRUCTIONS AS THE PC'S INSTRUMENT OF SUPERVISION

How the site manager processes the client/DT instructions can have a tremendous impact on the management of a project. Too many builders' standing instructions cover the mechanistic handling of incoming instructions, creating more paperwork in parcelling it out to subcontractors – almost sight unseen. These PCs having passed down an endless stream of unstudied instructions are genuinely surprised to receive on the rebound a flood of BS contractor's queries, perhaps accompanied by a slowdown in productivity as uncertainty takes hold, e.g. in response to a variation instruction for two shopping centre units to be made into one – 'Which of the two metered water and electricity supplies are to be omitted?' Maybe cellular offices are added and queries raised regarding the flexibility of HVAC units and 'Where do you want the light switches?'

Audited against the company standing instructions the site manager probably cannot be faulted for acting as a postman, but unless the SM understands what instructions he is passing on he will start to lose control of his project. Even if for speed he issued client/DT instructions under cover of his own PC instruction (PCI) he should follow it up as quickly as possible with either his own interpretation, or request an explanation from the BS contractor – 'What does this mean?' Yes, we understand how busy the site manager is and that on large jobs we are talking hundreds, if not thousands, of instructions. On those projects he is, or should be, supported by an organization that has written project procedures operated by people who can recognize important issues, particularly those instructions concerned with change, and keep them alive.

7.3.4 INSTRUCTIONS AS THE BS CONTRACTOR'S INSTRUMENT OF SUPERVISION

Through the BS contractor's organization there must be mirrored similar processes to those operated by the PC for the receipt and issue of instructions. The BS contractor must also avoid being a 'postman' by understanding or seeking explanation for the likely impact of instructions passed on to specialists and subtraders. The PC's site manager could request his company or site quality manger to audit this instruction handling aspect of the BS contractor's QP. With open lines and workable procedures for the receipt and implementation of instructions maintained 'live' by regular verbal communication, we should not lose sight of what is happening at the workface.

7.3.5 VARIATIONS

Touched upon briefly earlier, and to be considered again in Chapter 10, variations are so often mismanaged that it is no wonder they are the seed corn of conflict. One difficulty contracts have is in the 'what if' situation where the client DT is considering a change but wants to know what it will cost and whether it can be carried out without extension of time. For the site manager this is usually a no win situation. Next week his services contractor starts the lst fix ceiling level on the top floor offices and the last thing the site manager wants to do is issue him with drawings showing conversion to a management suite with higher spec finishes including English yew panelling. However secretive and sensible the site manager and BS contractor can be, the word will out and we enter the classic catch 22 situation. How can the intended variation be accurately priced and programmed, and the effect be assessed when, by the time the decision is taken, we can only guess where we might be with the current installation? Assessed it is, but the decision is slow in forthcoming and more work is actually installed by the time the formal variation instruction is received by the PC. Upon receipt of the variation order (VO) the BS contractor advises that his price and programme no longer hold and suggests daywork. The PC has some sympathy with his BS contractor and reports the situation to the client/DT. Relationships become strained, meanwhile work to the original spec and programme continues. The client considers he is being held to ransom and abandons the variation leaving the builder and services contractor with no means of recovering their expenditure. Worse still they are faced with a need to recover productivity due to the effect of rumour slowing the job down

There is no such thing as a typical variation. They are infinite in their variety and for this reason the site manager must always consider them seriously if the supervision of his project is not to slip away from him.

7.4 The process of inspection

7.4.1 GENERAL

Through the process of inspection the status of work changes from unapproved to approved. The importance of this should be burned into the mind of the site manager, for work of any volume, increasing over time, that has not been properly inspected is putting his project at risk. After all, whatever responsibilities others may have, it is the PC and the contractors with their specialists and subtraders who have undertaken to deliver the works on time, to the tendered cost, and to the specified requirements, aka quality.

The signposts along the way to understanding the process of inspection cover its resourcing, choosing what to inspect, and how to measure it through the management of defects. It is only through the acquisition of knowledge recorded through meaningful inspections that the site manager will obtain a true feel as to whether or not building services construction is achieving programmed targets with work of the right quality.

7.4.2 TYPES OF INSPECTION

For construction work there are two types of inspection:

- **progress**, which measures quantity; and
- **compliance**, which assesses quality.

If progress inspections are all about 'is it there?' compliance is to do with verifying that materials, commodities (plant, equipment and fittings) and the workmanship that turns them into BS systems, meets the specified requirements. In addition there is also a whole range of builders' compliance inspections which the site manager and his team will devise, sometimes with the BS contractor. These inspections will cover:

- builder's work marked out/completed;
- area acceptance;
- area completion;
- void closure.

Progress and compliance inspection come together at hold points. These may be specified, e.g. a water or gas main buried in the ground may not be back filled until it has been both inspected and pressure tested. The builder's compliance inspections listed above are also hold points on the progress route.

7.4.3 RESOURCING INSPECTIONS

The site manager must concern himself with the subcontractor's resourcing of his supervision and inspection and that provided by his team.

The resourcing of supervision and inspection will have been covered strategically at the pre-award meeting, under Item 3, organization (see Table 3.2). That skeletal framework will have been fleshed out through the provision by the BS contractor of quality, safety and environmental plans, particularly the QPs. Good QPs should list the verifying checklists that the BS contractor proposes to use. Now is the time to call up those forms. Ask the BS contractor to provide a list and samples of the forms he proposes to use to check the installation of the work. It is best to ask for a full set of forms first and follow this by requesting confirmation of the resources they will be using. If you ask for the latter first you may get a thinner level of resourcing than seeking appropriate 'full' inspection sheets which then have to be supported by inspection to discharge the responsibilities they impose. Doing it this way round is the better approach to securing competent inspection.

In a perfect world it should only be necessary for the PC to organize confirmatory progress inspections of the BS work in relation to preceding and following building trades, e.g. the area acceptance, completion and void closure aspects instanced earlier. These inspections are checks to ensure that the work in progress is being carried out to the approved method statements, particularly with respect to safe working practices and will form an important part of the site manager's activities. Increasingly the site manager will become experienced not only in those matters but also in the compliance and progress of the engineering work itself.

On jobs with small site teams the site manager may be sharing the inspection load with his general foreman. On larger jobs with commensurate bigger PC teams it is probable, and certainly recommended, that the level of supervision and inspection to be provided by a prospective BS contractor is checked before the order is placed. Once again the nature of services with their multidiscipline, multi-trade and specialist content comes into play. If at the pre-award/ management strategy meeting the BS contractor was examined on the level of supervision to be provided, at that time his response would have been partial. Until he has gone through the same examination of his specialists and subtraders that he was then facing from the PC he will be unable to make firm commitments as to what their supervision and inspection resourcing is going to be. With those sub subcontracts as yet unlet the BS contractor is initially only capable of giving a firm commitment with respect to basic mechanical and electrical systems. At best the PC can only record the BS contractor's own view as to what level of supervision and inspection the subtraders and specialist ought to provide, which they will be seeking. All of this means that as and when subtraders and specialist contracts are awarded and their plans for quality, safety and the environment are submitted, that is the time to investigate more fully the 'what, where, when and by whom' of supervision and inspection.

7.4.4 DIVISION OF RESPONSIBILITY FOR INSPECTION

This is not a repeat of 7.2 with its almost identical title. In that clause we looked primarily at supervision as the hierarchical conduit for the flow of instructions, some of which found their origins in inspection. There is a certainty that the client and DT representatives will be issuing instructions under a contract. There is less certainty about their roles in inspecting the works, and it is this that is given further examination.

A quick skip through the different terms of engagement for BS designers used in the industry gives rise to an 'optional service' feel about inspection. It is unwise for the site manager to make assumptions as to what level of inspection to expect from the client/DT. At the same time as he explores the division of responsibility for supervision (see section 7.2.1) he should find out what inspections the client/DT representatives will be making. There should be no surprise if upon raising questions it is found that greater clarity attends the right to issue instructions, but exposes some vagueness around planned inspections. Some of that vagueness will be removed if clerks-of-works are appointed. Their inspection lists carry with them the implied instruction for corrective action.

The site manager may find it helpful to create a matrix of responsibilities for supervision and inspection similar to the example in Table 7.1, which ties in three levels of control for QA, management, and supervision and inspection on a management contract. Under the terms of the contract and in the preliminaries and preambles of bills and specifications client/DT representatives will have set out some of their assigned duties for inspection of information delivered by the BS contractor, e.g. in relation to working drawings, QPs, safety and environmental plans, etc. These will receive comment, if not outright approval. The inspection of samples and mock ups submitted or constructed before the 'in place' construction occurs are usually clearly defined hold points for Client/DT approval. This applies also to offsite manufacture and testing.

The BS designer's responsibilities for inspection on site could be as Clause 2.8(f) in Appendix 1 of the ACE Conditions of Engagement, 1995, Agreement B(2):

> Attend relevant Site Meetings and make other periodic visits to the Site as appropriate to the stage of construction or as otherwise agreed to assist the Lead Consultant to *monitor that the works are being executed generally in accordance with the contract documents and with good engineering practice* and advise the Lead Consultant on the need for instructions to Contractors. The frequency of Site Meetings and periodic visits by the Consulting Engineer shall be as specified in The Memorandum of Agreement or as otherwise agreed between the Client and the Consulting Engineer.

Table 7.1 Responsibilities matrix – supervision and inspection for a management contract

Aspect	Control	1st lead	Aspect matters addressed	2nd lead	Aspect matters addressed	3rd lead	Aspect matters addressed
Quality asurance (QA)	Overall	DT/PC	Project quality strategy PC implements by QP	PC internal audits	QP subjects	Extrinsic audits	QP subjects
Quality assurance (QA)	Package	Works contractor	QA (management) to meet tender award requirements	PC external audits	Audits works contractor QP or site QA system	Extrinsic audits	Audits WC QP or site QA systems
Management	Overall	PC	All planning, organization, coordination, control	DT	Arising from contract review meeting (report)	Project principals	As referred by DT/PC
Management	Package	Works contractor	All planning, organization, coordination, control	PC	Through regular package review meetings	DT	Arising from contract review (report)
Supervision	Overall	PC	Instructions to works contractor	DT	Instructions to PC	Visiting agencies	According to their responsibilities, e.g. fire. safety, building control
Supervision	Package	Works contractor	Instructions to workforce	PC	Instructions to works contractor	DT	Instructions to PC
Inspection (QC)	Overall	PC	Provides levels of QC verification to agreed project quality strategy	DT	Instructions to works contractor (via PC)	–	–
Inspection (QC)	Package	Works contractor	Provides QC verification to agreed levels	PC	Confirms WC QC verification to levels agreed in PC QP	DT	Confirms WC QC verification to levels agreed in PC QP

Note: Where DT specify they require to be represented to witness tests they will be the 1st lead in that aspect of the package works.

What you get with that is 'a look' by the designer when he is on site. The 'look' will be sharper if it has been focused by the designer's site staff who may be a BS clerk-of-works. The value of these 'looks' by visiting designers can be variable and questionable in their contribution to the management of the project. For the small BS job, say up to £0.25 million located no more than one hour's travel from the representative designer's office or home, and with Site Meetings starting mid-morning, the designer, limited by appointment to a monitoring role, might just make a useful contribution through the observations made on a walk round the job prior to the site meeting commencing. Appointed with the same monitoring role to a larger job or even the small job which is located so far away from office or home base the designer's contribution may be of negative value:

1. Unable to take a pre-meeting 'look' the designer may feel he has got to say something and raises matters that he thinks are relevant, but because he is out of touch do nothing but waste time.
2. The designer may be inadequately represented. The fee bidding by multi-service designers precluding the attendance at site meetings of adequate representation for all BS disciplines. The attending representative is asked questions which are answered inappropriately, inaccurately, or time is lost in their referral 'to the office'.
3. After the site meeting the DT representative takes a look at the job and tries to raise matters of concern with anyone who will speak to him, not always observing the contractual routes, before making tracks for home. The difficulty the DT representative faces in trying to get to see the architect or site manager is that they are tied up in the usual structured sub-meetings and inspections that have arisen out of the site meeting held earlier in the day. The services designer returns to the office and possibly up to the time of the next site meeting creates a misconceived air of efficiency by telephone, fax and letter as the issues of the site meeting and 'look around' are addressed.

The above criticism of the design engineer should not be directed at them but those clients, cost consultants and PMs whose concern is lowest cost rather than the value they may get from money spent on purchasing meaningful inspections. Years ago when buildings and services technology were simpler, the 'look around' visit by an experienced multi-service engineer was of value. The advances of technology embodied in widening ranges of systems, many microprocessor controlled, more closely integrated with building structure and fabric and needing to comply with care based safety and environmental legislation, have diminished the value of the 'look around'. Applying their talents on a quick look around in the spotlight of today's stage even the best multi-service engineer is 'hanging by the fingertips'. Any inspector can only take to the job that combination of knowledge and experience of his personal track record. Some inspectors recoil at the enormity of what is expected of them, others accept the challenge, are of a questioning mind and can sort and concentrate on the key issues. These comments are made as guidance for the site manager who having discovered the assigned roles for inspection learns to judge and trust those that bring reliable information.

7.4.5 INSPECTION SHEETS

The biggest question the site manager has to ask of the BS contractor is, 'You are responsible for installing the Works to the specified requirements. How are you going to verify progressively that this is being achieved?' When the BS contractor responds with his proposed forms, checklists or whatever other name the inspection sheets may be graced with, their adequacy should be checked. It is important that inspection sheets should be appropriate so that the content of what is checked truly contributes to the avoidance of defects. Whether it is a design engineer's duty or not, it is suggested that it is in the interest of the PC and BS Contractor to reach agreement on inspection sheet content so that 'the job is adequately inspected and our assigned responsibilities properly discharged.'

BS contractors have inspection sheets that range from collections to properly organized libraries. Those branded as collections are assorted sheets from old, recent and perhaps current projects. Where they form part of, or support, a QMS system these collections are becoming better organized. Often they are a hotchpotch of ill considered content with little standardization about them. Other firms have better libraries with standard inspection forms for a range of plant, equipment and distribution systems. These firms may also have standard sheets that they expect their subtraders and specialists to use.

The content of what should be inspected needs to be appropriate and to consider the function of the building, spaces being served and whether or not the BS are exposed or enclosed. Wherever services remain exposed and are reasonably accessible for maintenance and repair, the detail of the inspection sheet may be less onerous than if those services were enclosed and work on them later would be disruptive. Of course, there are exceptions to the rules, for example where reliability is paramount as in a continuous process factory or laboratory experiment, or where the repair of services within a prison building attracts consequential costs for access and attendance upon the repairer. Wherever building services are installed in health care installations above local GP surgery level, or their failure brings immediate threat to life such as the release of pyrofluoric chemicals in wafer fabrication plants, then the most stringent inspections (and associated tests) are warranted. These definitions bring within their orbit domestic installations such as electrical and gas services. There is, in fact, very little that falls outside the need for meaningful inspection.

Some examples of inspection sheets associated with the general distribution of building services and lifts are included for guidance in Appendix J. The piping, ducting and insulation sheets are reproduced with the kind permission of Haden Young Ltd. The reader is also directed to the inspection sheets contained in the NALM quality plan of their publication *Principles of Planning and Programming a Lift Installation* [3]. For a job specific example refer to Fig. 5.6(a) and (b) where N.G. Bailey related their inspections to the content of 1st, 2nd and 3rd fix activities. Electrical installations must be inspected to verify compliance with BS7671, 1992. Requirements for electrical installations, formerly these requirements were embodied in the IEE Regulations. There exists a comprehensive body of published work available in the form of guides from the IEE, the Association of Supervisors and Executive Engineers and others, containing recommendations on inspection, much of which goes hand in hand with testing.

There will be occasions where the BS contractor proposes the use of inspection sheets of inadequate content, nullifying their usefulness. The judgement on their adequacy may come from the DT or the PC. On these occasions the DT/PC may be able to offer suitable sheets for adoption by the BS contractor. In these instances the PC should insist that the BS contractor adopt the inspection sheets as his own. Ownership is important. The BS contractor must not be relieved of his responsibility for providing work that is compliant with the specified requirements when measured against an inspection sheet provided by others.

The BS industry is in continual change, for example in the application of tungsten inert gas welding of pipework, the on-site assembly of sheet metal ducting, and in the finishing of insulation. All of these bring improvements in productivity enabling the work to be done right first time, quickly. This rate of progress can quickly date inspection sheets and serves to emphasize that what to inspect requires careful job specific assessment.

7.4.6 WHAT TO INSPECT

The site manager cannot know all that the client and DT representatives will inspect, but he can get the BS contractor to summarize from the specification what the client/DT have said they *are* going to inspect.

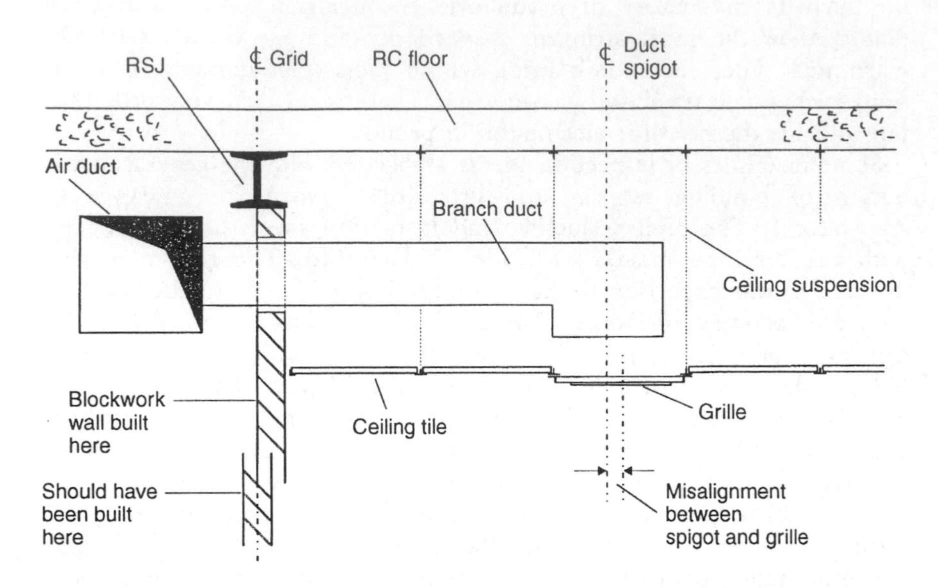

Figure 7.1 Who is to blame – the bricky, the ducter, the ceiling tile erector?

What the site manager should be interested in having properly inspected is BS work at the interface with building works. Wherever building services attach to, pass through the structure and fabric of a building and are fixed into or on its surfaces, they should be inspected. It is at these interfaces that the site manager will need to have both preceding and following building trade work inspected:

- Services fixed to the structure must be inspected to ensure that structural integrity has not been degraded, e.g. removal of fire spray on steelwork.
- Services passing through structure and building fabric boundaries of fire compartmentation will be required to be fire stopped.
- Services passing through structure and building fabric not forming fire compartments may nevertheless require smoke barriers and acoustical stopping.
- Services fixed into false floor and false ceiling membranes need checking for position, alignment, sealing and stopping.
- Services fixed on building fabric surfaces, e.g. masonry walls, partitions and on fixed pattern surfaces such as tiling, need checking for location, coordination and alignment.

The site manager, in the first instance, may be looking for all of this to be inspected and verified by the BS inspectors. In a sequence of first, second and final fix BS activities interlaced with building trades, it isn't all down to the services firm. The situation in Fig. 7.1 occurs so frequently it must be a 'classic'. Sheet metal duct work manufactured to engineering tolerance off gridline dimensions leaves a spigot turned down to serve a grille in the position dimensioned on the architect's reflected ceiling plan. The masonry wall built in the wrong position but to the usual building tolerances is not discovered until after the false ceiling grids have gone in. The ceiling firm having used initiative rather than querying the difference between the drawing and what exists on site. Now we are at the point of discovery and it is cheaper for the ducting firm to get out the tin snips and alter the sheet metal than it is for the defaulting companies to correct their work. Summarizing the situation, the services installer should inspect *all* of his works and where they interface with other trades. For the builder inspections must be concentrated on the interfaces.

7.4.7 WHEN TO INSPECT

Work should be inspected, before, during and after the activity. At the interface of building and services work this means:

- confirming a work area is suitable for BS to commence;
- the BS contractor inspecting the work during the activity and
- confirming to the builder that the work is complete and
- commencing another defined activity in the same area or
- releasing the area back to the builder for following trades;
- the work of trades as they follow that left by the BS contractor should be inspected to confirm the achievement of the interface, without damage.

The site manager should arrange spot check inspections of the first piece of work that is carried out by each new BS trade that is commenced on site. This has a twofold benefit; it notifies the BS contractor that you are serious about inspections and should ensure the achievement of specified standards from the beginning. We all know how much easier it is to lower a high standard than it is to raise a standard that is unacceptably low.

The initial spot check should start as the material arrives on site:

- Is the drainage material as spec?
- Is the sheet metal ducting of the right gauge?
- Is the steel piping the correct weight (medium or heavy), mild steel or galvanized?
- Is the cable on the drum the same as the drum markings?

That's got the general M & E underway but it is just as important when the new specialists and subtraders arrive later in the job, to carry out the same check on the first activities of their work.

The site manager should arrange inspection for the access routes to be clear for the delivery of plant and equipment into the building. It is not uncommon for plant to arrive and find that access which was clear 24 hours earlier is now blocked by sand, ballast, a water barrel and concrete mixer. Do not release an area then through the activity of building work render the space unusable. Having reached agreement with the BS contractor that riser shafts are available for the installation of distribution systems you will hear soon enough if the temporary brick edgings at upper level have been removed and the shafts are being used as refuse shutes. It is your project and you, the site manager, should be inspecting that the BS contractor is working to his method statements' safe working arrangements and not being put at risk by the unsafe practices of others.

Returning to the subject of interface inspections and in order to avoid conflict, arrange for inspections to take place as one contractor vacates a work area and is followed in by another. Certainly BS contractors have damaged the work of others; but in return they have suffered greatly. Under many subcontracts they are responsible for the protection of their works in areas that they are not working in, and the nature of their work makes it unprotectable, e.g. all the services above the false ceiling in an air conditioned, sprinkler protected office building are at risk as ceiling contractors fix primary and secondary grids. In turn, the ceiling contractor's work becomes at risk as wiring is pulled through and final connections made from AC terminals to grilles and pre-wired light fittings plugged in. By making worthwhile inspections at these interfaces the site manager can aid the development of a 'caring site' culture.

7.4.8 DEFECTS

In an imperfect world defects will arise from a lack of understanding of what is required and a shortfall in inspection by the person doing the work. As that person moves along on the same piece of work or commences a new activity believing it to be compliant with requirements, he enters the domain of the inspector. But didn't we define inspection as an examination that could also take place during an activity? We did, but during an activity the supervisor may issue a verbal instruction instigating immediate corrective action in order to avoid a recordable defect being noticed by others. This is a not uncommon cycle of combined supervision and inspection at the workface by the workforce's management. It results in unrecorded rework by the BS contractor which the PC has difficulty in spotting unless it is obvious, e.g. prefabricated boiler header pipework being observed in position one day and a few days later being 'nowhere on site', having being returned to the prefab shop. Observations of current incorrect construction may be recorded by the DT clerk-of-works and notified to the site manager. From this example we can see the importance of the site manager agreeing with the BS contractor that all defects are to be recorded whether they arise from inspections during or after completion of activities. Both parties must come to an agreement as to how work is to be inspected, defects recorded and cleared.

The site manager should ask for copies of all the BS contractor's own recorded defects and reinspection or 'defect cleared' sheets and logs. This information will repay valuable study from which the site manager must judge:

- Are the defects considered to be excessive?
- How long after the raising of a defect does it take for the clearance sheet to come through?
- Despite the defects and rework is the BS contractor still on programme?

If the answer to the last question is no, it raises questions about the impact on the following trades and risk of damage attendant upon the BS rework. Answers to these questions and his own views repay study time put in by the site manager and empower him to act. All of this valuable knowledge has come about by the site manager making sure he armed himself with the results of the BS contractor's own inspections. It is those inspections that should be the most expert and regular. Possibly, and some may say not unreasonably, the BS contractor may be reluctant to provide details of his inspections because he has not carried out enough and/or the general level of workmanship is not up to standard. In these situations the site manager becomes involved in an extractive process that is nevertheless worth pursuing. Now is the time for the site manager to look for leverage within the BS contractor's QP and any agreements made with him for his QMS to be audited by the PC. The route to information on a reluctant BS contractor's defect management status is through a reminder of the promises made through his QP, approved method statement and any right established by the PC to audit them. BS contractors may have ulterior motives and cut back on inspection hoping that others will tell them if their installation work is not up to standard. Although the PC may only be left with the DT/clerk-of-works and his own inspections of the BS contractor's first work on site and at the interfaces with other trades mentioned earlier, he should avoid increasing his level of inspections. As a last resort where a BS contractor defaults on verifying compliant work in voids, then the PC may wish to notify the contractor that 'arrangements will be made to have the work in voids verified before closure, the costs of which will be to BS contractor's account'. Regrettably, we are deep into conflict of which difficulty over establishing compliant work may be but another manifestation of a poorly performing contractor. Every PC with a QMS will have a standard sheet for advising of a non-conformance. This sheet may be used for other purposes. Intended primarily for use in notifying work not complying with specified requirements it can also be used in association with BWIC and temporary works, e.g. electrics etc. To maintain the benefits of a recognizable format the non-comformancies identified by the DT/clerk-of-works or any of the external inspectorate parties could be attached to this sheet.

The great emphasis of this section on inspection has been on compliance with specified requirements; but what about progress to time?

References

[1] Harding, Colin (1994) Time for architects to relinquish their leadership, *Building*, 11 February.

[2] Torrance, Jack (1987) The building site minefield' *Building services*, February.

[3] NALM (1994) *Principles of Planning and Programming a Lift Installation, with relevance to Escalators*, London.

Assessing construction progress 8

8.1 General

All site managers should feel comfortable with the best practices for controlling work. They were succinctly set out by G.J. Pinder in the passage 'Subcontracted Specialists, Construction Management in Principle and Practice' [1]:

> The basis of any form of control is a realistic target and performance measured against that target, enabling any deviation in performance to be scrutinised and corrective action taken, where necessary. Planning procedures are the normal means by which targets are set, and regular progressing enables deviation from the programme to be examined. If the specialists concerned have been a party to preparing the targets and supplying supporting data in respect of the progress made, then they are obligated to contributing towards any necessary solutions which are required to bring a project back on to programme. This form of control tends to be automatic and more acceptable, and leads to more cooperation. Trying to exercise control without the means of targets is fruitless.

Planning and programming are the most important of all management functions. Projects whose construction programmes depict all the services activities in three or four bar chart lines are certain losers as their simplistic approach provides no basis against which progress can be measured. Projects where programming has been given its rightful place as the key management function should have a foundation for proper progress measurement. Unfortunately, the shortfall in programming may only be discovered by a site manager in seeking confirmation of his fear that the BS contractor is slipping. Some site managers 'freeze' and do nothing, or accept the BS contractor's 'stories' that they are 'on programme'; 'These activities are always a bit slow, but we'll catch up later', 'I'll put another pair on next week.' Other site managers may sensibly seek support from higher up the BS contractor's line management, or their own, but lose time hesitating in making the case for a planning resource that has not been costed. If they get the resource, which in the circumstances needs to be the best, they may have to make do with what is available. Meanwhile the project spirals downwards.

The better planned projects with their hierarchy of targeting tools –

- design programmes
- BS contractor award schedules
- master programme
- managing programme
- sectional programmes
- short-term and/or stage programmes
- weekly programmes
- commissioning programme

provide a firm foundation for progress assessment.

It is only on a properly planned project that the site manager will be able to use knowledge of the BS contractor's activity content, method, supervision and inspection in the measurement of progress. That knowledge, coupled with good programming enables the SM to run regular progress meetings with the BS contractor, acquire and evaluate the records of production output and take effective action to keep on target.

Where even the best managed PCs get into difficulties is in the number of BS contractors and specialists that are involved. The more BS contractors and specialists directly employed by the PC, the greater the managerial workload in controlling their works. The great advantage here is the direct control this affords the PC, especially in programming. The PC is fortunate if the contract allows him to package the BS works to his and the project's advantage. The client/DT, recognizing that PCs are not always capable of doing this, often dictate the way in which services are packaged. So, we may have a multi-service ME & P package which includes a number of specialist subtraders, say BMS, security and fire detection and prevention (see Tables 4.1 and 4.2). Alternatively each of these may be separate BS contractors. If the PC is responsible for the way in which the BS works are packaged and he chooses the multi-service route then he assigns the detailed responsibility for the control of the works to that contractor, while retaining overall responsibility for managing their integration with the building works. If, to avoid profit upon profit attendant upon sublet works he decides to opt for a greater number of single BS work packages, then the PC retains a higher level of responsibility for overall and detailed control of those works. In this instance the PC must recognize that any saving he makes on the capital cost to him is eroded by the increase in managerial resource that is required to control more contractors. The process of control and assessment of progress that follows is one that must be replicated either by the PC or the BS contractor for each services contractor or specialist employed.

8.2 The site progress meeting

8.2.1 AGENDAS IN GENERAL

For the site manager, a simple objective of any meeting held on his site, for whatever purpose, must be to aid and record the path of progress.

We will consider here the site progress meeting agenda for a BS contractor who may be of single discipline, e.g. public health, multi-service or specialist such as a catering equipment supplier and installer, all in the direct employ of the PC. For all of these there is a core to the agenda. This must address the key issues of:

- safety
- quality control
- environment
- drawing approvals
- information and/or decisions required or outstanding
- progress of offsite production
- progress on site
- variations.

This core shortlist will vary according to the geography of the job and the scope of work of the BS contractor, be he general or specialist. So far nothing in these general terms differentiates the BS agenda from any other element of work. By now the site manager is acclimatized to the difference in BS work and will not be surprised to find that the Agenda needs to be 'topped and tailed' to make it appropriate to the start up of a BS contract with its delivery of information programming – see section 5.2 – or to the change in emphasis for commissioning and handover, to be dealt with in Chapters 9 and 11. The next sections create two sample agendas. The first, for the front end of a BS contract, covers the situation up until construction is in full flow, and the second takes us down through commissioning and handover. It would be too daunting to have an agenda which covered building services from the beginning to the end, although from Chapters 9 and 11 the site manager will perceive the need to keep commissioning and handover in his mind at all stages. For it is the site manager, in the absence of a lead from the BS contractor, who must decide the change in emphasis from the first agenda to the second.

8.2.2 AGENDA FROM START ON SITE TO MID-CONSTRUCTION

Integrating the core agenda into one that will take us from start on site to mid-construction we arrive at Table 8.1. The differences between building and services elements begin to show and will become even more prominent as the agenda is customized to take account of site geography and services features. Having recommended the site manager to hold an information production planning meeting he is now advised to use that agenda, Table 5.1, again as items 1.1 to 1.7 in the site progress meeting. By doing this the site manager is retaining a continuous firm managerial hold on his project. Too many site managers go adrift between award

Table 8.1 Agenda for site progress meeting from start on site to mid-construction

1. Report on information
    1.1 Quality, safety and environmental plans
    1.2 BWIC drawings
    1.3 Working drawings
    1.4 Method statements
    1.5 Sample approvals
    1.6 Test and inspection plans (construction)
    1.7 Procurement schedules
2. Information and/or decisions required from DT
3. Progress of procurement and offsite production
4. Progress on site
5. Inspection, construction testing and defect clearance
6. Variations

and start on site. Not until three or four progress meetings have passed does it start to dawn on them that the information by which the services should be installed is not flowing through as it should. Without stamping his authority on these issues with the use of a good agenda the site manager will find himself talking to the BS contractor in terms of recovery programmes for the delivery of information. Shortly, the same exercise will need to be carried out on the real work.

There are no clear boundaries to define what scope should be covered under any one meeting item. Under information and/or decisions required from the DT it may be found that some matters have already been brought up under item 1, e.g. the approval of drawings and samples taking the DT longer to grant than under the agreed project procedure. It is under item 2 that requests for information (RFI) and technical query sheets (TQS) are reviewed. No one wishes to see meetings drag on unnecessarily and these reviews of information often need to be no more than the numerical status of the mailing system, everyone appearing pleased when able to record that there are few outstanding matters. It is worth the site manager throwing in a question or two to see how important the RFIs/TQSs really are. Why not ask 'Do any of the RFI/TQS sheets answered since our last meeting affect progress of information production or work on/off site?' Beware the BS contractor who seems to be raising a lot of unnecessary TQSs; they have been known to be a smoke screen for not producing drawings, as the contractor strings out the drafting process.

Gradually, after the last BS contractor or specialist has started on site, the volume of information to be delivered, and its status to be reviewed at site meetings, will diminish. With it production should rise to a plateau.

Item 3, the progress of offsite production, may also have been covered earlier as a sub item of 1, sample approvals, or again in test and inspection plan review. Unless a visit to the manufacturer has occurred recently or is about to take place then information concerning status may be vague. Without feedback from a recent visit and the next one being some way off the site manager may request the BS contractor to ring the manufacturer for a verbal update to be included as a post-meeting note. At item 4, progress on site, we have reached the point that most site managers consider to be the sole purpose of the site meeting. Discussions about progress at the workface can quickly degrade into differences of opinion about what has been done and what needs to be done, with little or no evidence being produced by either side. Section 8.3, reporting methods, offers guidance.

If item 4 has measured the quantification of progress then item 5 looks at quality under inspection and defect clearance. Despite being inseparable in theory, they can be looked at differently in practice. Drawings marked up to show lots of installed pipework, ducting and conduit are heartening and can be financially beneficial to the PC and BS contractor when included in the monthly valuations. When welds subsequently fail specified tests, air ducts leak at an unacceptable rate and conduits have no draw wires or have been cast in the wrong position, the job is in trouble; the necessary rework does nothing for progress along marked up bar charts or coloured services plans. No new work takes place as resources drop back on remedials. If inspection and defect clearance closely shadow the installation work, recorded progress is meaningful and work has not been 'overbilled'.

There is a tendency to look only at recent variations (item 6 of the agenda), and not the whole picture. Now and then it is worth taking a broader view to consider the rate and value of work accruing from variations. The hard times of the 1990s made PCs and BS contractors more commercially astute in managing variations; unfortunately this is not matched by control of their impact at the workface. So the site manager should keep them under review, for many small changes can attract drafting and office engineering costs in a way that far outweighs the work content, causing drawings to go round another approval cycle and end up with work at tender estimate rates being subeconomic. Perhaps it is because the very word 'variation' is an emotive one in construction that the impact upon programmed work is not properly evaluated. When you say you want to 'review the variations' the BS contractor comes to a meeting with his commercial team. He is disappointed you aren't talking money. You're disappointed he's not talking progress. The site manager must set out the ground rules. There is usually a contractual right for instructions to be issued which vary the work. The same contractual obligation extends in most cases to the recipient having to accept such instruction. Therefore the work that was originally contracted for now has to be built with the variation content in it. It is that revised content which must be reviewed to ensure that the resources of plant, equipment, material and labour are provided to meet contractual handover.

8.2.3 AGENDA FROM MID-CONSTRUCTION TO HANDOVER

Unlike football, construction is not a game of two equal halves and the exact point at which the site manager changes the agenda for his BS site progress meeting needs to be chosen carefully. As the need for the delivery of information fades away after the appointment of the last BS contractor or specialist and those appointed nearer the beginning of the project approach the mid-point of their construction, this could be the point to change the emphasis to 'handover'.

In the agenda of core issues, safety appeared at the top and it will not have gone unnoticed that it does not appear as a main item on the agenda of Table 8.1. It should, however, have featured prominently in the subitems of 1.1, quality, safety and environmental plans and again under 1.4, method statements. Here in the agenda of Table 8.2 it is also covered in two areas. As the first item on the agenda to remind the BS contractor of site wide safety arrangement meetings and special training, etc. Safety will also be featured in item 4, production progress, which must be to safe working methods. The comments for items 2–6 are as for the same numbered clauses in Table 8.1.

Remember that these are your meetings; you call the shots, you set the agenda and because of this the BS contractor will rarely request to talk about commissioning until it is too late for both of you. As the site manager you must be proactive and raise the subject first. You can

Table 8.2 Agenda for site progress meeting from mid-construction to handover

1. Safety
2. Information and/or decisions required from DT
3. Progress of procurement and offsite production
4. Progress on site
5. Inspection, construction testing and defect clearance
6. Variations
7. Commissioning
8. Documentation (manuals and record drawings)
9. Training and instruction programme
10. Final inspection and clearance of defects
11. Handover

always 'try it on': 'Do you think at the next meeting we should put commissioning on the agenda and talk about the preparation of your method statements for system flushing and cleaning, pre-commissioning, balancing, etc., etc?'. On a large project the designer may have specified in the preliminaries the requirements for managing commissioning through a separate series of meetings. If not, in response to your question the BS contractor may suggest the separate approach. For the smaller project for say heating and ventilation, low voltage power and lighting, simple security, fire detection and protection, usual public health, i.e. a simple job with no process services, then commissioning may be dealt with at the general progress meeting. Even though it is simple the job will go through the same process as covered in Chapter 9. Judging the point at which to change the emphasis from construction to commissioning is not easy, but it is better to start early than late. For guidance the author suggests that on contracts of overall duration of nine months commissioning should be put on the agenda halfway through that main contract period. For jobs between 9 and 15 months' duration change the agenda eight weeks before commissioning is due to start. Getting to the bigger league, 18 months to two years and beyond, the consideration of commissioning on your progress meeting agenda should be four to six months before any related activity is due to start on site. Jobs over nine months to a year in duration will almost certainly benefit from separate commissioning meetings held just before the general site progress meetings, which then need only summarize the progress of commissioning. The advice on timing the introduction of the subject of documentation for BS in the form of manuals and record drawings follows the recommendations above for commissioning (see also Chapter 11).

Item 9, training and instruction programme, must also be considered along the same timescale as the commissioning and documentation. The key consideration here is that the end user has competent staff available who can be given specified training and instruction. What you don't want is a DT on the instruction of the client, or the client, trawling your job looking for excuses for not taking it over because they have not appointed anyone to run it. The subject of training and instruction is covered in Chapter 11.

The active site manager will have thought ahead and considered the preparation of his project for the final inspection and clearance of defects process. The inactive site manager will stroll into a minefield. The message is to set your stall out and get your BS contractor thinking ahead with you to resource inspections and clearance of defects well ahead. You will get a more comfortable ride from the client/ DT who find their workloads reduced through your preparation. Guiding detail on this is provided in Chapter 10.

The agendas in Tables 8.1 and 8.2 provide a framework that must be covered. Inform the BS contractor at the beginning of the job that these will be your agendas, reserving the right to change them as necessary. They should not be rigid or preclude the setting up of special site meetings, or devoting a particular site meeting to a knotty construction problem. Listen to the BS contractor and make your own judgement as to the merit of their suggestions, always being wary of motives he may have for diverting the way in which you are directing progress.

8.3 Reporting methods

8.3.1 GENERAL

To accurately fix the position of BS progress at a moment in project time is difficult. Precise assessment is costly for providers. By comparison, the superficial and volumetric measurement of most building work elements makes assessing their progress much easier. Preliminaries clauses of BS specification and some PC's own requirements for BS contractors issued at tender stage make attempts to define what must be provided for the assessment of progress. More usually this is left to be sorted out at site level between the PC and BS contractor.

Once again it is down to the site manager, perhaps aided by his visiting QS if it is a small job, or supported by a site QS and planning resource on larger projects. The PC and BS contractor will come to an agreement on the format and detail with which the latter presents their progress report. The degree of confidence the site manager will have in that report depends on the depth of meaningful detail he has been able to get the BS contractor to provide. Bear in mind that in the negotiation the BS contractor will have started with the attitude of providing as little as possible, as the provision of information costs money, and the site manager will be looking for the greatest amount of reassuring documentation he can get. What is provided is a compromise and this is why it is important for the site manager not to take the services progress information at face value, passing it for valuation and writing it up in the monthly report for the Client/DT. The PC, particularly the site manager should form their own opinions of the BS contractor's progress. In doing this the site manager, and his team if it is a larger project, will bring to bear their knowledge of:

- method of construction
- range of activities
- content of activities: first, second and final fix
- duration of activities
- status of construction testing.

Remember that by now on a well managed project the 'aids to assessing progress' exist in the project files in the form of:

- quality, safety and environmental plans
- programmes
- inspection and defect records
- status of witnessing.

Using these aids firm up your own views, which in the end become subjective even when based on hard evidence. It is worth considering what other information is available in the form of reports, comments, and notifications that have been received from the client/DT, the external inspectorate of regional utility companies and local authority inspectors, between the last progress review and the one now under consideration.

The most valuable discussions on site progress are those where the information on progress is provided prior to the meeting. This allows time for the recommended research on its validity to be undertaken. It is recognized that if meetings are held at weekly intervals the pre-supply of information may be impossible but at fortnightly periods it should be achievable. Certainly for the monthly valuation, the basis for information provided by the BS contractor in terms of progress, for which he is claiming payment, should be discussed at the preceding site progress meeting. It is amazing how much overbilling occurs that the site manager becomes aware of too late. The benefits of enhanced cash flow viewed encouragingly at head office drain the site manager's cost leverage at the back of the job when there is much performance to be extracted from the BS contractor. Commissioning and preparation of documentation can easily cost 3–4% of the services job value for which the services firm finds it has spent those funds on installing the work. The fools' paradise created by overbilling is a land not only inhabited by the constructors, but one also visited by the design team where their fees are related to the valuation of progress.

Two suggested BS contractors' reports to be submitted to the PC prior to a site progress meeting are included in Appendix K. The first report in Appendix K is for use at the front 'half' of the job and is compatible with the agenda in Table 8.1. The second report in Appendix K is suitable for the second 'half' and relates to the agenda of Table 8.2. Referred to here for completeness relative to reporting methods, it covers much that will be more comprehensible after reading Chapters 9, 10 and 11.

8.3.2 THE DETAIL OF REPORTING METHODS

In support of their subcontract report BS contractors should provide information on progress:

- marked up programmes and/or
- progress status sheets.

To keep the information in digestible form the best method is considered to be percentage completion of an agreed list of activities related to the programme. Unless the activities correlate to the programmes, evaluation of the percentages becomes difficult. The BS contractor who earlier provided a paucity of programming information but who now suddenly provides copious lists of activities with percentages against them which cannot be quantified is probably heading down the route to overbilling. In construction progress reporting the use of status sheets can be most helpful on jobs with small repeated spaces such as hotels, prisons, hospitals, apartments and office units. An example of a M & E suite installation status sheet used on a multi-storey development is shown in Table 8.3. Each of the 250 suites comprised an office of approximately 100 m^2, self-contained with toilets and kitchenette. The 'suite' layout showing the four-pipe fan coil HVAC system is on Fig. 8.1 to be read in conjunction with the status schedule. It is only by summating the individual status reports on such cellular projects that overall predictions of percentage completion have any meaning. Ferreting around in rabbit warrens will only give rise to empty 'gut' feelings that are a waste of energy.

8.3.3 MONITORING PROGRESS AND EVALUATING REPORTS

Monitoring progress commenced in Chapter 7; the recorded output can now be used to evaluate the progress reports. By using information in his possession –

- defect clearance sheets from the same source, together with
- signed, witnessed construction testing sheets and
- area release and void closure sheets.

the site manager can spot check to ensure that the claimed progress is to standard. If it is not then the evaluation and progress records should be written down and the BS contractor called upon to improve on what should be a rolling programme of inspection and defect clearance.

8.4 If it goes wrong

What do you do if it goes wrong? Regrettably it will, for every construction project is a prototype. In this prototyping situation the theories of design solutions and the advances of technology are called upon to be proven. We are back to the importance of assessing risks to the construction of the project from the technology it involves. Even with the best appraisal and management to mitigate risk we will face the unknown. Being unknown it will be unexpected when problems occur. That puts framing thoughts to the risks of engineering technology. But even the best-managed jobs may quickly run off course for other

Table 8.3 Mechanical and electrical 'suite' installation

PRODUCED BY
R.S. GRICE/P.E. JEFFREY

DATE: 20.05.91 BLOCK: EAST
FLOOR: 7TH

		1ST FIX					2ND FIX							
SUITE No	SYSTEM	MAIN INSTALL	MECH TEST	LAGGED	ELEC T'MNATE	CEILING RELEASE	FLEXS & GRILLES	FIRE FITTINGS	SANITARY WARE	SANITARY FITTINGS	ELEC T'MNATE	ELEC TEST	SNAGS COMPLETE	
7.5	APARTMENT SUPPLY	100	100	100	N/A	100%		N/A	N/A	N/A	N/A	N/A		
7.5	APARTMENT EXTRACT	100	100	N/A	N/A	100%		N/A	N/A	N/A	N/A	N/A		
7.5	TOILET SUPPLY	100	100	100	N/A	100%		N/A	N/A	N/A	N/A	N/A		
7.5	TOILET EXTRACT	100	100	N/A	N/A	100%		N/A	N/A	N/A	N/A	N/A		S
7.5	KITCHEN EXTRACT	100	100	N/A	N/A	100%		N/A	N/A	N/A	N/A	N/A		U
7.5	FAN COIL UNIT	100	N/A	N/A	N/A	100%		N/A	N/A	N/A	N/A	N/A		I
7.5	LPHW TO FAN COILS	100	100	100	N/A	100%	N/A	N/A	N/A	N/A	N/A	N/A		T
7.5	CHW TO FAN COILS	100	100	100	N/A	100%	N/A	N/A	N/A	N/A	N/A	N/A		E
7.5	APARTMENT SPRINKLERS	100		N/A	N/A	50%	N/A		N/A	N/A	N/A	N/A		
7.5	TOILET/KITCHEN SPRINKLERS	100		N/A	N/A	50%	N/A		N/A	N/A	N/A	N/A		
7.5	H/LVL ELECTRICS APARTMENTS	100	N/A	N/A	N/A	100%	N/A	N/A	N/A	N/A				C
7.5	H/LVL ELECTRICS TO FCU'S	80	N/A	N/A		40%	N/A	N/A	N/A	N/A	N/A			O
7.5	H/LVL BMS WIRING	100	N/A	N/A	100	100%	N/A	N/A	N/A	N/A				M
7.5	H/LVL FIRE ALARM WIRING	100	N/A	N/A	100	100%	N/A	N/A	N/A	N/A	N/A			P
7.5	H/LVL P.A. WIRING	100	N/A	N/A	N/A	100%	N/A	N/A	N/A	N/A				L
7.5	L/LVL ELECTRICS APARTMENTS	100	N/A	N/A	100	100%	N/A	N/A	N/A	N/A	N/A			E
7.5	HWS PIPEWORK TO TOILET	100	100	100	N/A	N/A	N/A	N/A			N/A	N/A		T
7.5	CWS PIPEWORK TO TOILET	100	100	100	N/A	N/A	N/A	N/A			N/A	N/A		E
7.5	HWS PIPEWORK TO KITCHEN	100	100	100	N/A	N/A	N/A	N/A			N/A	N/A		
7.5	CWM PIPEWORK TO KITCHEN	100	100	100	N/A	N/A	N/A	N/A			N/A	N/A		
7.5	TOILET DRAINAGE	100	100	N/A	N/A	N/A	N/A	N/A			N/A	N/A		
7.5	KITCHEN DRAINAGE	100	100	N/A	N/A	N/A	N/A	N/A			N/A	N/A		
	TOTALS	99%	87%	100%	75%	90%	0%	0%	0%	0%	0%	0%	0%	44%

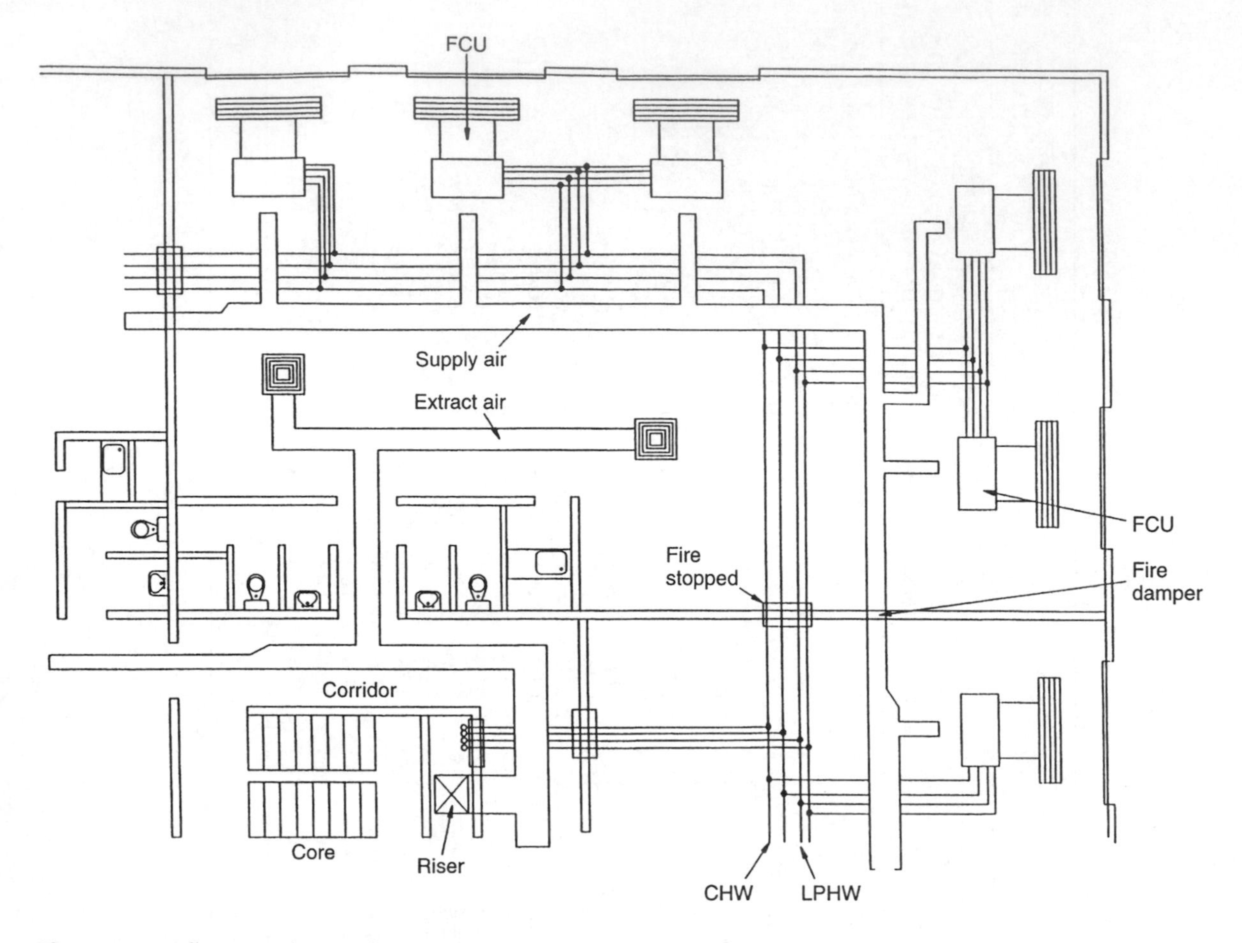

Figure 8.1 Office suite layout showing four-pipe fan coil HVAC system.

reasons. Perhaps it would have been better to have entitled this section 'when it goes wrong'; but we'll stick with the more optimistic 'if it goes wrong', and plan our action to establish:

- effect
- cause
- how it is to be remedied.

8.4.1 INSTALLER

Sticking to the subject of construction progress or lack of it, as the problem, it might be best to first establish the cause. No doubt the first discussions will take place during a site progress meeting in which the installer is examined to check:

- that his work is commencing on time and in the right location;
- if labour resources are adequate;
- if material stores are adequate;
- if plant and equipment has been delivered;
- that construction plant and tools are adequate;
- that timekeeping is satisfactory;
- that supervision and inspection levels are appropriate.

The discussions of the meeting should be verified by a walk round the site. The subjects of the above list are resolvable, but until the situation is recovered the site manager should establish the effect of performance shortfall with respect to:

- quality
- quantity
- other trades.

Cause and effect now having been established the PC and BS contractor can sit down and agree proposals for recovery which may include:

- changing the labour;
- increasing the resources of material, labour, construction plant and tools;
- increasing and/or improving the level of supervision, inspection and defects clearance.

Pouring additional resources into the job may not be an efficient way to correct defaulting progress. Recovery must be planned and monitored against the use of short-term programmes. The BS contractor should be advised of the knock on effect and consequential costs that will be directed to his account by the PC when received from following trades who have become delayed. Naturally, the situation is reversed when building trades cause delay to the BS contractor through failure to complete their works on time.

8.4.2 DESIGN

Where design is the cause of delay to the BS installer it may be for two reasons: fit and performance. Both are areas of risk which may have eluded the PC's appraisal of the building services. The first of these, fit, will be dealt with here, leaving the failure of system performance to be discovered later, in the commissioning process.

By far the largest amount of BS design work is done on the basis of a specification and drawings. It is the design drawings that are the source of the problem. When a designer says he is providing a fully designed job there is plenty of scope for misunderstanding. This problem and the proposed solution for the industry is well defined in BSRIA's Technical Note TN8/94 *The Allocation of Design Responsibilities for Building Engineering Services – a code of conduct to avoid conflict* [2], to which the author contributed. Unless the designer has embraced the recommendations of that document it is most likely that in saying he has produced a detailed design, to which no further description is added, the PC and BS contractor are led to believe that they are getting a higher level of detail than that actually provided. In these situations the designer produces drawn information along the following lines:

- a set of general arrangement drawings, say 1:100 scale;
- a set of schematics (this is a bonus);
- single line plant room layouts (possibly 1:50 scale);
- a few coordination details of obvious 'pinch' points where services leave plant rooms (at their largest size); difficult crossovers; typical services risers including toilet services duct;
- plant and equipment schedules (these may be included in the specification or on the drawings).

All services design is a numerical solution the fit of which has to be proven through the working drawings and installation. Herein lies the difficulty. Take a simple case of radiators below windows, running between columns that protrude into the occupied space. The designer, who has after all declared that he has prepared a full design, may reach agreement that the general arrangement drawings can be used, at the installer's risk, as the working drawings. On site the radiators are found not to fit. The designer walks away from the problem, the PC reschedules the work and the claim starts to be built up against the BS contractor. The BS contractor obtains shorter, higher, fatter radiators. At the end of the job to the dismay of the client, an incoming tenant tries to negotiate a reduction in rent due to the effective 'net lettable area' not being provided by the protrusion of the fatter radiators into the space. Who was at fault? The BS installer attempted to meet specified requirements by putting scheduled radiators in the space shown on the drawings. Surely he was right. The designer's case is that if the BS contractor had produced working drawings, as specified, this problem would have been discovered and resolved. He thinks he is right. We will leave this problem of 'fit' to be resolved by sensible compromise rather than the court, and deal with the subject of 'system performance' in the next chapter.

8.5 The effect of variations

It is usual for PCs to have standing instructions covering the receipt and logging of variations. Sometimes variations have to be raised by the PC and BS contractor, as they are discovered 'buried' in all the general information (drawings, schedules, answers to RFIs and TQSs, etc.) emanating from the DT, and issued under an architect's instruction (AI). This system puts the onus on the PC and his contractor to study the information issued by the DT for differences from that tendered upon. The first instance of this is the receipt of the DT's contract issue drawings. The differences spotted are summarized and a request for a variation instruction made out. The PC must also find out the power of the clerk-of-works to issue instructions. For these and any verbal instructions the PC will have procedures and forms for acknowledgement of site instruction and confirmation of verbal instruction. The PC's site instruction procedures generally run back to back with the BS contractor's. Variations therefore fall into two categories: those that are clearly instructed and those that have to be raised by the installers as a result of discovered differences between the theory of the tender and the practice of construction.

8.5.1 DESIGN CHANGES TO DRAWINGS ISSUED FOR APPROVAL

Between the two clear categories of variations referred to above there can lie a disruptive grey area that the PC will become involved in. Regrettably, some designers do indulge themselves and make changes to their designs through the comments they make on the installer's working drawings issued for approval en route to acquiring construction issue status. The range of changes may be along the lines of:

- adding or omitting a fitting, e.g. sink unit;
- changing sizes on distribution system, e.g. piping, ducting cable tray, etc.;
- changing the position from which terminals are served, e.g. from one riser to another, or from low level to served from high level below.

Many of these changes will be seen to be minor arising from 'preference engineering' – 'I like to see it run this way.' Much that falls into this grey area means little by way of change in the work content. It is in the drafting, engineering, commercial and administration time that the costs, and they are considerable, truly lie. A drawing issued for approval can be returned with so many minor changes causing no impact to the work content, but requiring an awful lot of drawing work and resubmission around the approval cycle. On large projects, if this starts to happen, the PC is recommended to get together with the BS contractor and set their case out clearly and early for if it is allowed to continue, delay in productivity will become alarming.

Variations on technical aspects are observed to occur on a project in the two halves that can be associated with the agendas of Tables 8.1

Table 8.4 Variation data sheet

CONTRACT: ______________ Addendum type: Variation Order/Letter/Site Instruction/Drawing/Claim item/C.V.I.*	*Delete as applicable
Reference No: ____________	Date Received: __/__/__
Had work to be varied already commenced?	
Did the order refer to completely new work?	
Did the order cause the stoppage of work by any operatives – if so how many for how long?	No. of tradesmen: No. of labourers: Time of stoppage:
Did the work take longer because of this instruction?	
Did this order stop any plant? Did this order cause any change in plant requirements?	
Did the order cause the re-direction of operatives – if so, to where?	Re-directed to: Distance: metres Tradesmen: time lost Labourers: time lost
Did the order cause the re-direction of plant, if so, to where?	
Did this work disrupt the work of any sub-contractors?	Name: No. of men: Time disrupted:
If the variation affected sub-contractors please state sub-contract order numbers.	

Table 8.4 *Continued*

Did this order require the re-ordering of material? If so:	Staff:
• How long did re-ordering take to requisition?	Hours:
• Were previously ordered materials made surplus?	Type:
How were surplus materials disposed of?	
Delivery date for new materials	
Estimate of staff time taken in discussion, meetings, research, phone calls due to this variation	
Does this order require a notice to the Engineer/Architect in accordance with: • ICE Clause 12 51 • GC/Wks/1 Clause 7 9 • JCT '63 Clause 11 24 • JCT '80 Clause 13 26	Please give data & reference no. of such notice
Does this order require notice for claiming extension of time? • ICE Clause 44 • GC/Wks/1 Clause 28 • JCT '63 Clause 23 • JCT '80 Clause 25	Please give date & reference no. of such notice
Any other information:	
Form completed by:	Signed:

and 8.2. During the preparation of the BS contractor's working drawings and the building of the systems TQSS will be raised, the responses to which can give rise to a variation instruction. Over that first half the variation orders (VOs) will be very few compared to the second half. In this period, as the building with its services approaches its final form and it starts to appear to the client and DT that in three dimensions it is not what was quite envisaged on the drawing board or on the CAD screen, so the VOs start to flow. Coupled with the built works, variations attributable to the stages of commissioning, e.g. system preparation, pre-commissioning, regulation, control, validation of the BMS and structured cabling, etc., etc. start to rise on an exponential curve. At a time when a site is working at the fine interfaces of finishes and services terminals and everyone is getting excited about how the project is looking, or should look, we are treated to some very indigestible fare. Under these circumstances it is easy to understand why the control of so many jobs is lost. It is not the content of the variations but the sheer number that cause the problem in even the best managed teams. While drafting in more commercial aid gets a grip on the money angle, it isn't so easy to achieve the essential levels of communication, resourcing and constantly updating the programmes that has to be done with many subcontractors. The benefit of having a multi-service BS service contractor is apparent; but only if he too can handle the volume of variations. There is no easy answer to this problem. If the site manager calls for more supervision and planning help, will he be able to recover their cost in the variations? Unlikely, and the same will apply to the multi-service contractor who is also being overrun. What can happen is that the site manager and his contractors muddle along until with some clarity it is obvious that the job is going to overrun on time and cost, and quality is beginning to suffer. The author believes it unfair to expect the contractors to solve the problems of others in an ever-compressing timescale with liquidated and ascertained damages hurtling towards them. A plea is lodged for those at the front end of the project process to discharge the responsibilities of their function and not burden the contractors with their shortcomings.

To end on a practical note PCs may find the variation data sheet included as Table 8.4 of value in assessing the impact of variations on construction progress. Fine for variations arriving at a steady flow, but daunting to complete and evaluate in a waterfall.

References

[1] Pinder, G.J. (1971) *Subcontracted Specialists: Construction Management in Principle and Practice*, Longman.

[2] BSRIA (1994) *The Allocation of Design Responsibilities for Building Services – a code of conduct to avoid conflict*, Technical Note 8/94, Bracknell.

Commissioning and its management 9

9.1 General

9.1.1 COMMISSIONING MANAGEMENT

The management of commissioning oversees a change in emphasis from construction to the delivery of a functioning project. It is not a road to be travelled hopefully in the time remaining after construction, but a journey that must be planned, organized, coordinated and controlled. It is a journey to be completed on time and for this there needs to be a strategy. For the PC's site management team the strategy for arriving on time is simple. Using the handover date they should produce a backwards-looking commissioning programme. The strategic management programme of construction in Fig. 3.1, expanded with commissioning detail, becomes Fig. 9.1, our backwards-looking strategy for the management process. As the backwards-looking commissioning programme meets forward moving construction there will be clashes to be overcome and a carefully studied order of priority needs to be established.

It was stated in BSRIA's Technical Note 17/92 *Design Information Flow* [1], that

> All services systems must be commissioned and the Designer should have no difficulty in specifying the general technical references, standards, codes and guides for each service. However, the Designer will need to think each system through and be specific in identifying technical requirements.

It is normal for the activities of commissioning to be the contractual responsibility of the building services installer i.e. the installer must delegate or carry out in house, such activities as system cleaning, pre-commissioning checks, setting to work, proportional balancing and system testing etc. The installers will also have the responsibility for their own management of these 'hands on' activities. Where the services are extensive, complex and carried out by a number of contractors and their sub-traders, the proliferation of managerial elements concerning commissioning cannot be left to them. Hence the need on complex jobs for an integrating commissioning management function.

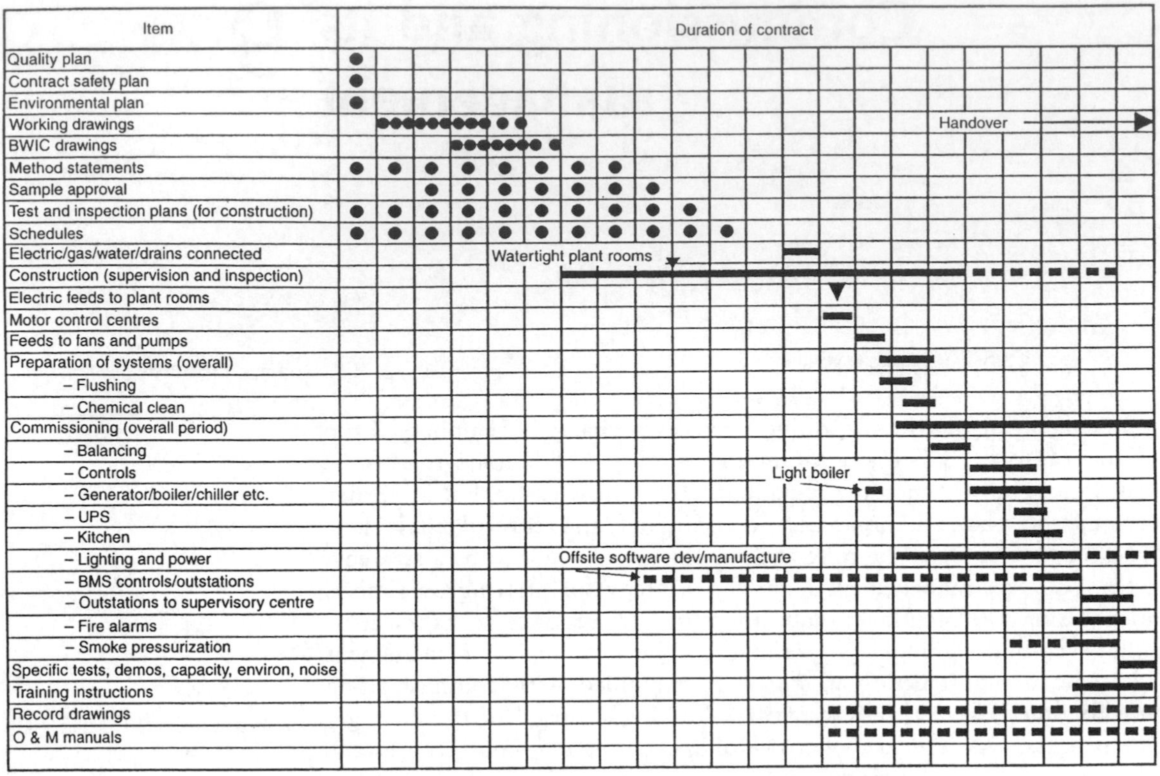

Figure 9.1 Strategic management programme with commissioning detail.

In truth commissioning must be managed on all jobs and the site manager must acquire an understanding not only of commissioning and its management, but also the impact of those activities on:

- the required condition of building work elements
- safety
- finishing trades

These aspects will brought out in the following sections:

- the commissioning management specification;
- planning and programming;
- the process of commissioning;
- post-contract – system proving and fine tuning;
- recorded output of the process.

As an aid to understanding some definitions would seem appropriate

9.1.2 DEFINITIONS

The definitions of commissioning and its management in Table 9.1 are drawn from the BSRIA's Application Guides published in the late 1980s. The author contributed to these definitions as a member of the steering committee for the production of the guides. The CIBSE use similar terminology in their Commissioning Codes.

It is important for the site manager to note that the commissioning specialist is usually the firm carrying out the hands-on commissioning activities. There may be a number of these on any project, for plant and equipment such as boilers, chillers and generators, and for systems such as HVAC. The project's specification may call for the appointment of a commissioning manager with overall responsibility for the planning, organization, coordination and control of all commissioning activities. This appointment may be a firm independent of any of the commissioning specialists. Alternatively, particularly in the case of a multi-service BS contract, that firm may be called upon to provide a commissioning manager from within its own resources. On many jobs the site manager will find that the commissioning management process is not clearly defined. In these instances, and they may be the smaller projects, the requirements for commissioning management will be found dispersed within the contract conditions, preliminaries and preambles, and specification clauses such as those listed in Table 9.2. Even where a particular commissioning management specification does exist it is well worth a trawl through the contract documentation to ensure that there are no conflicting or precedential statements. It is recommended that the site manager asks the BS contractors to collate all of the commissioning requirements specified in their contract and provide him with a copy.

9.2 The commissioning management specification

9.2.1 GENERAL

The commissioning specification sets out the duties of an independent commissioning management firm or calls for the assignment of a specialist manager from within the multi-service or lead BS contractor. A good specification will also define the roles and responsibilities of the client and DT.

9.2.2 RANGE OF DUTIES

Drawn from the author's *Design Information Flow*, BSRIA TN17/92 [1], Table 9.3 lists the range of duties for commissioning management. The duties may vary according to the timing of the appointment of a commissioning manager. Sometimes, usually on the larger projects, the specification may include a division of responsibilities schedule.

Table 9.1 The definitions of commissioning for HVAC, BSRIA AG 2/89.1

Commissioning	The advancement of an installation from the stage of static completion to working order to specified requirements
Commissioning specialist	The firm or person appointed to carry out specified duties in connection with commissioning the engineering services
Designer	The firm or person appointed to design the HVAC systems
Commissioning management	The planning, organisation, coordination and control of commissioning
Commissioning specification	The document that prescribes in detail the requirements with which the commissioning service has to comply Note: Specification must refer to drawings, schedules and the relevant parts of codes, manuals, guides and other standards
Commissionable systems	Systems designed, installed and prepared to specified requirements in such a manner as to enable commissioning to be carried out
System	A heating, ventilation or air conditioning concept of equipment, distribution ducts, pipes and terminals, associated or independent so as to form a complex unit
Installation	A system placed in position
Environmental performance criteria	The specified, numerically quantifiable, characteristics and tolerances of an internal environment to be achieved by the HVAC system
Design criteria	Those measurements and quantities selected as the basis for the design of a system
Testing	The measurement and recording of specified quantifiable characteristics of an installation or parts thereof Note: This also includes offsite testing
Pressure and leakage testing	The measurement and recording of pressure retention, and fluid losses or gains in the plant, equipment, distribution ways and terminals
Flushing	The washing out of an installation with water to a specified procedure in order to remove manufacturing and construction detritus
Chemical cleaning	The internal treatment of a pipework installation with chemical fluids to specified requirements
Static completion	The state of the system, installed in accordance with the specification, clean and ready for setting to work. In the case of water systems this includes flushing, cleaning, filling and venting

Table 9.1 *Continued*

Pre-commissioning checks	Specified systematic checking of a completed installation to establish its suitability for commissioning
Setting to work	The process of setting a static system into motion
Regulation	The process of adjusting the rates of fluid flow in a distribution system to achieve specified values
Environmental testing	Measurement and recording of internal environmental conditions
System proving	Measuring, recording, evaluating and reporting on the seasonal performance of the systems against their design values
Fine tuning	The adjustment of a system where usage and system proving has shown such a need. This may also include the re-assessment of design values and control set points to achieve the required system performance

9.2.3 DIVISION OF RESPONSIBILITIES SCHEDULES

As a proponent of the use of responsibility schedules the author wrote, for BSRIA, *Commissioning HVAC Systems: Division of Responsibilities*, published by them as Technical Memorandum 1/88 [2]. Appendix A of the TM contains a 24-item schedule of activities, from taking the brief to handover and beyond, with defining responsibilities for:

- employer
- DT
- managing contractor
- HVAC installation package contractor
- commissioning specialist.

Clause 8 of that document commented:

> Most projects would benefit from a clear understanding of the Division of Responsibilities. Whilst the Schedule has been specifically drawn up for Commissioning HVAC works, its principles can be applied to all building services, including those where commissioning is carried out by the installer.

The key aspects of any division of responsibilities schedule are the hold points it defines. TM1/88 listed the provision of reports as hold

Table 9.2 The dispersed requirements for commissioning management commonly found within contract conditions, preliminaries and preambles, and specification clauses

Commissioning (and testing)
Flushing
Chemical cleaning
Water treatment
Offsite testing
BMS (and controls)
System proving (acceptance tests)
Fine tuning
Programming (logic diagrams/networks)
Method statements
Emergency/standby mode operation
Training and instruction
Witnessing (and demonstration)
Approvals (including statutory, utilities and insurance bodies)
Certification
O & Ms
Record drawings
Handover (phasing, and beneficial occupation)
Practical completion

points recording the state of HVAC at:

- post installation
- systems cleanliness (after flushing)
- pre-commissioning
- commissioning (after setting to work and regulation)
- testing (of environmental conditions and capacity)
- system proving
- fine tuning.

As chair of the subcommittee for the steering group that guided the production of BSRIA's Technical Note TN8/94 *The Allocation of Design Responsibilities for Building Engineering Services – a code of conduct to avoid conflict* [3], the author led the preparation of item 4 of Appendix B, specifying system commissioning activities – see Appendix L. These pro formas take us from the HVAC specifics of TM 1/88 to the general and identify 28 activities, of which 24 refer to the management of commissioning. These general purpose schedules may be used for the management of any individual BS or, as primarily intended, for the integrated management of all the BS of a project. Site managers can expect to find these forms or others of similar principle forming part of a project's commissioning management specification.

Table 9.3 The range of duties for commissioning management, BSRIA, TN 17/92

Liaisons with designers
Obtaining design information
Agreeing commissioning methods and tolerances
Obtaining services specifications, drawings, standards, codes and guides
Carrying out design appraisals
Liaison with main, managing and services contractors planning departments
Producing with contractors input, coordinated commissioning logic networks, methods and programmes
Commenting on working drawings
Agreeing works test procedures and witnessing
Attending construction management meetings
Approving contractors' flushing and cleaning method statements and witnessing results
Preparing stage reports (see BSRIA TM1/88)
Providing a watching brief during installation
Witnessing pre-commissioning activities
Agreeing pro forma documentation
Witnessing and collating on- and offsite pressure testing
Establishing start-up and operating procedures
Checking test instrumentation
Witnessing test results
Establishing emergency procedures and witnessing plant close down
Defining system handover protocols

It is possible that even on quite large projects the requirements for commissioning and its management are distributed throughout the contract documentation – see Table 9.2. For these types of projects it is recommended that the site manager asks each BS contractor or specialist in the contract to prepare their own activity and responsibility schedules. In the absence of an assigned commissioning manager the PC has to integrate the management of the schedules into a complete process through planning and programming.

9.3 Planning and programming commissioning

9.3.1 GENERAL

The definition of commissioning as 'the advancement of an installation from the stage of static completion to working order to specified requirements' is time related and will include activities described as system preparation and pre-commissioning at the beginning of the programme, and system proving and fine tuning at the end. The last two activities have advantages for all if the nature of the project allows them to be specified to be carried out post contract. Views on this will be developed a little later.

Such is the optimistic 'can do' psyche of the human mind that every chiller, boiler and generator manufacturer installing his own equipment is confident in the answers they give to the question, 'How long will it take you to commission it?' Every time, the answer is given on the rarely stated premise of 'Given full access, and availability of gas, water and electrical power, etc., and all other necessary dependencies to be provided immediately.' Even the most experienced are prone to overoptimism based on this 'if everyone helps me as they should, I'll get it done in the time I said.' Unfortunately 'everyone else' also has a lot to do in providing utilities, completing construction, preparing systems and pre-commissioning them. They too could commission their systems if no one else got in the way. The purpose of a commissioning programme is to take the individual plant, equipment and system logical sequences and integrate them from individual programmes into one optimum duration commissioning programme. It is not easy.

En route to producing the overall commissioning programme clashes of priority may be discovered, service to service and service to finishing trades. The site manager must take control to ensure that as the backwards looking programme for commissioning meets the oncoming construction and finishing programme any clashes of priority are resolved. Guidance was given in section 8.2.3 on the use of the agenda in Table 8.2 and its associated contractors' report in Appendix K as to when to introduce the subject of commissioning into the site progress meetings.

9.3.2 LOGIC DIAGRAMS

Each major item of BS plant or equipment has either an unavoidable or preferred logic for its commissioning. This also applies to the BS system that the plant/equipment serves. It is easier for the plant and equipment manufacturer to be accurate in stating the time required for commissioning its product than it is for the BS contractor to be finite for the system. The former is usually dealing with a piece of proven equipment while the latter, although he has dealt with many similar systems previously, is dealing with the geography of the one in question for the first time. The degree of completeness of the system, possibly affected by other trades, contributes to the difficulty of assessing time. But assessments of activity durations are made and block by block for plant and system, logic diagrams become resourced programmes. Figures 9.2–9.5 are examples of logic networks.

Figures 9.2 to 9.4 show examples of plant, equipment and system logic diagrams that in very complex projects become 'blocks', as for Fig. 9.5. On that diagram it will be noted how a number of system groups

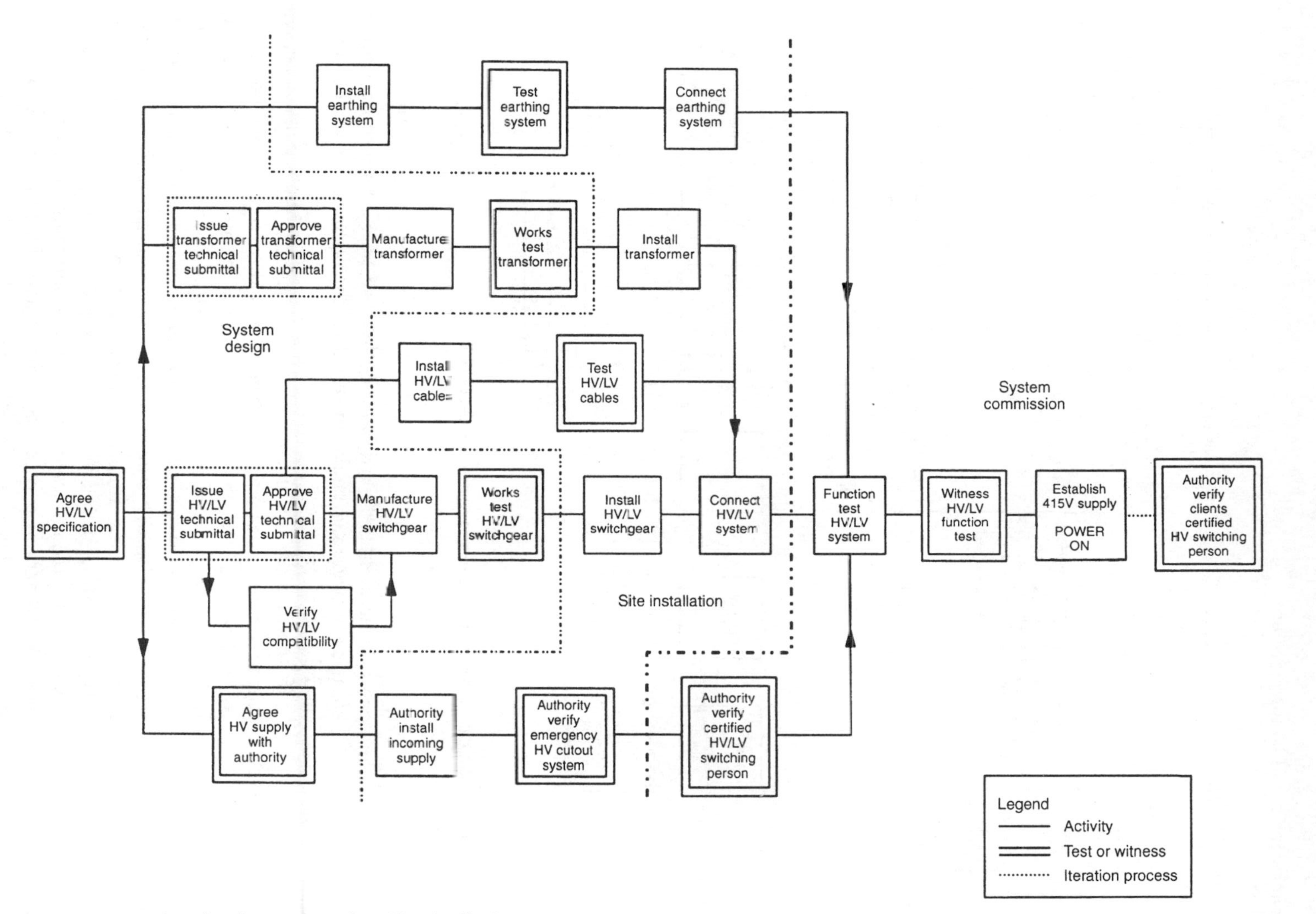

Figure 9.2 HV/LV development and verification logic.

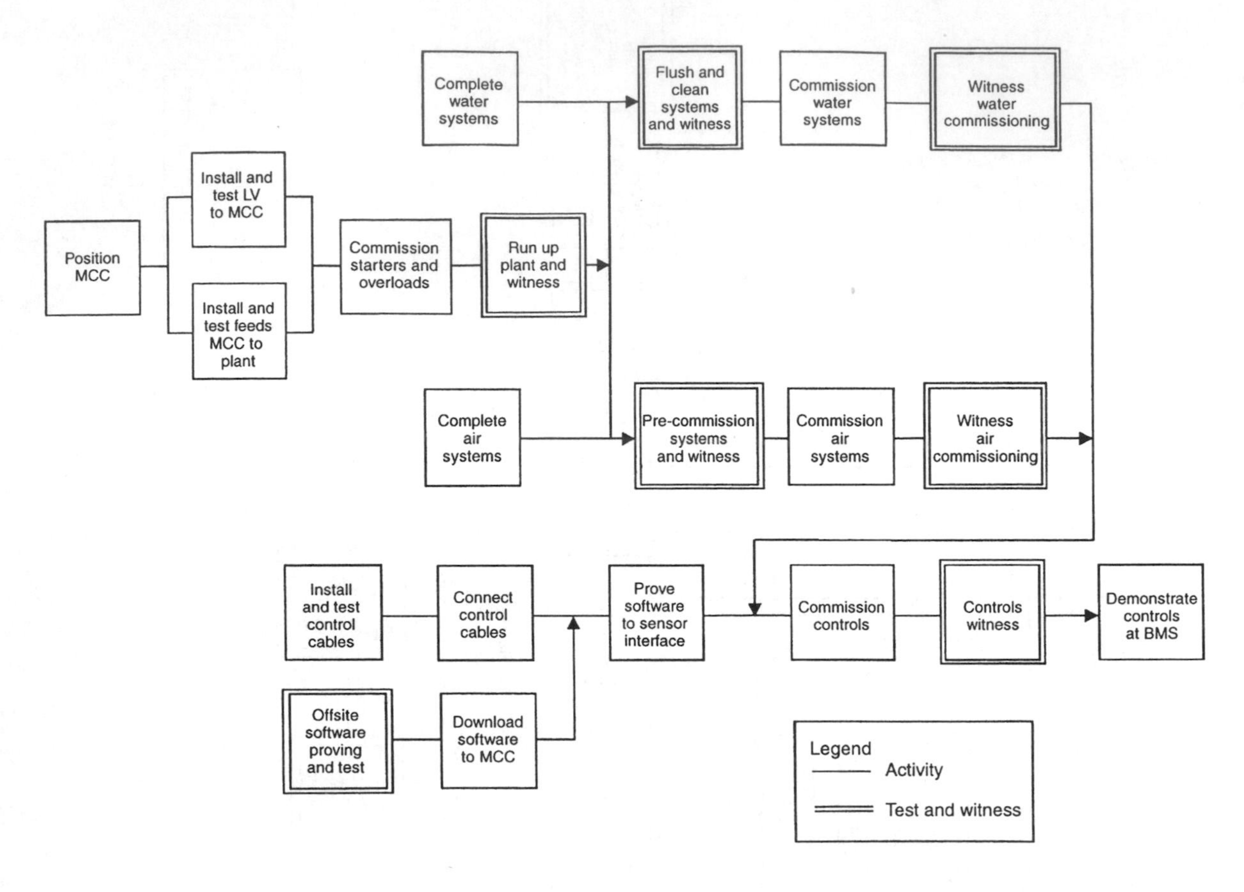

Figure 9.3 H & V mechanical/motor control centre typical completion logic.

can be commissioned partly in parallel, and how all are so dependent on electrical power; and finally in series go through the eye of the needle in modal operation.

If the project documentation has assigned the appointment of a commissioning management firm, or called upon the multi-service contractor or a lead BS contractor to appoint a commissioning manager with overall responsibility, the site manager is fortunate. If not, he has either to extract that resource from the multi-service BS contractor or at worst undertake the coordinating role in order to create an overall commissioning programme. This can develop into an area of conflict as each BS contractor and specialist makes the point that their tender was based on the economics of doing the commissioning according to their own logic diagrams and related programme. This may not be true in the sense that in order to tender it was not necessary for that contractor to

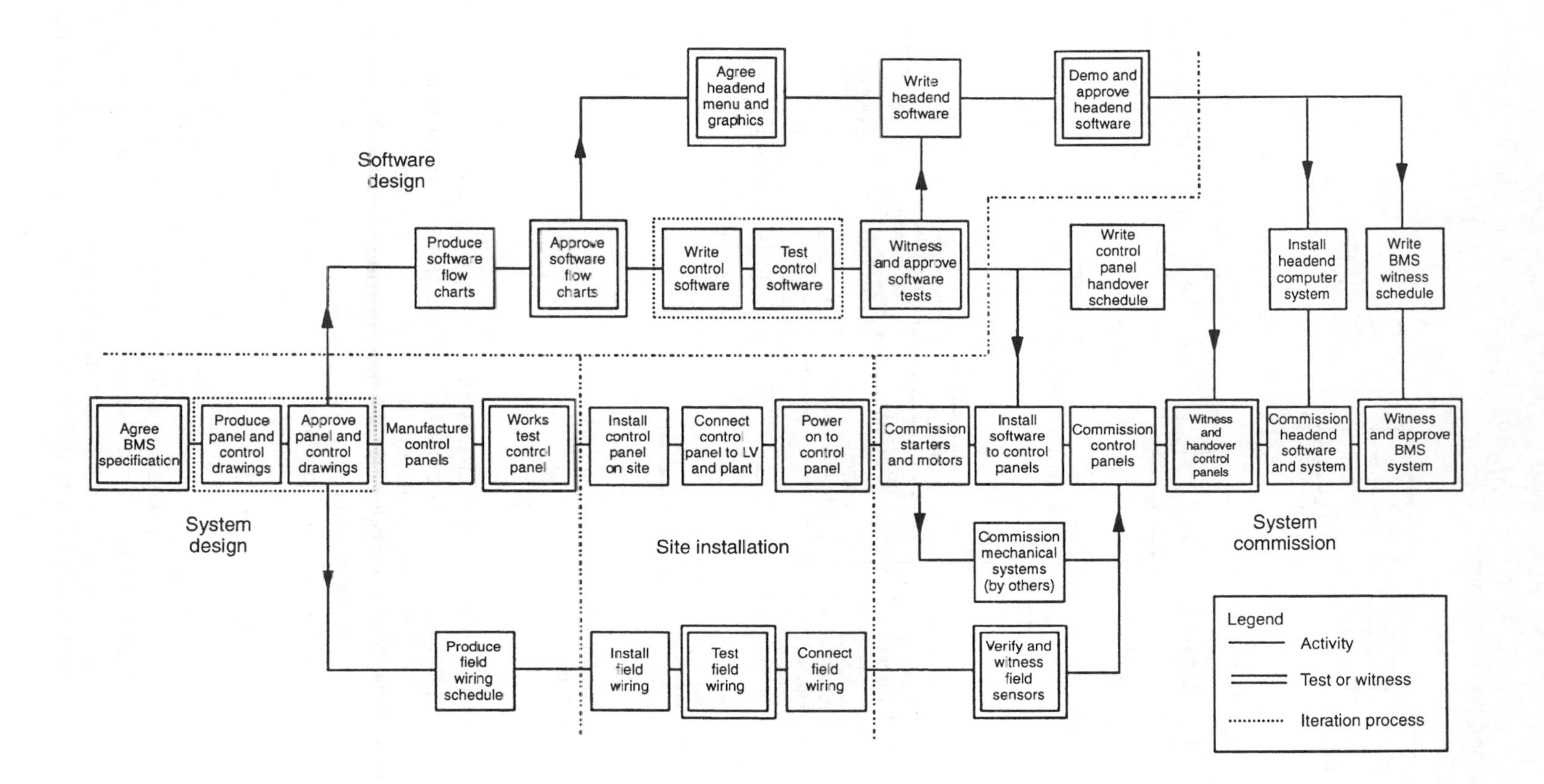

Figure 9.4 BMS development and verification logic.

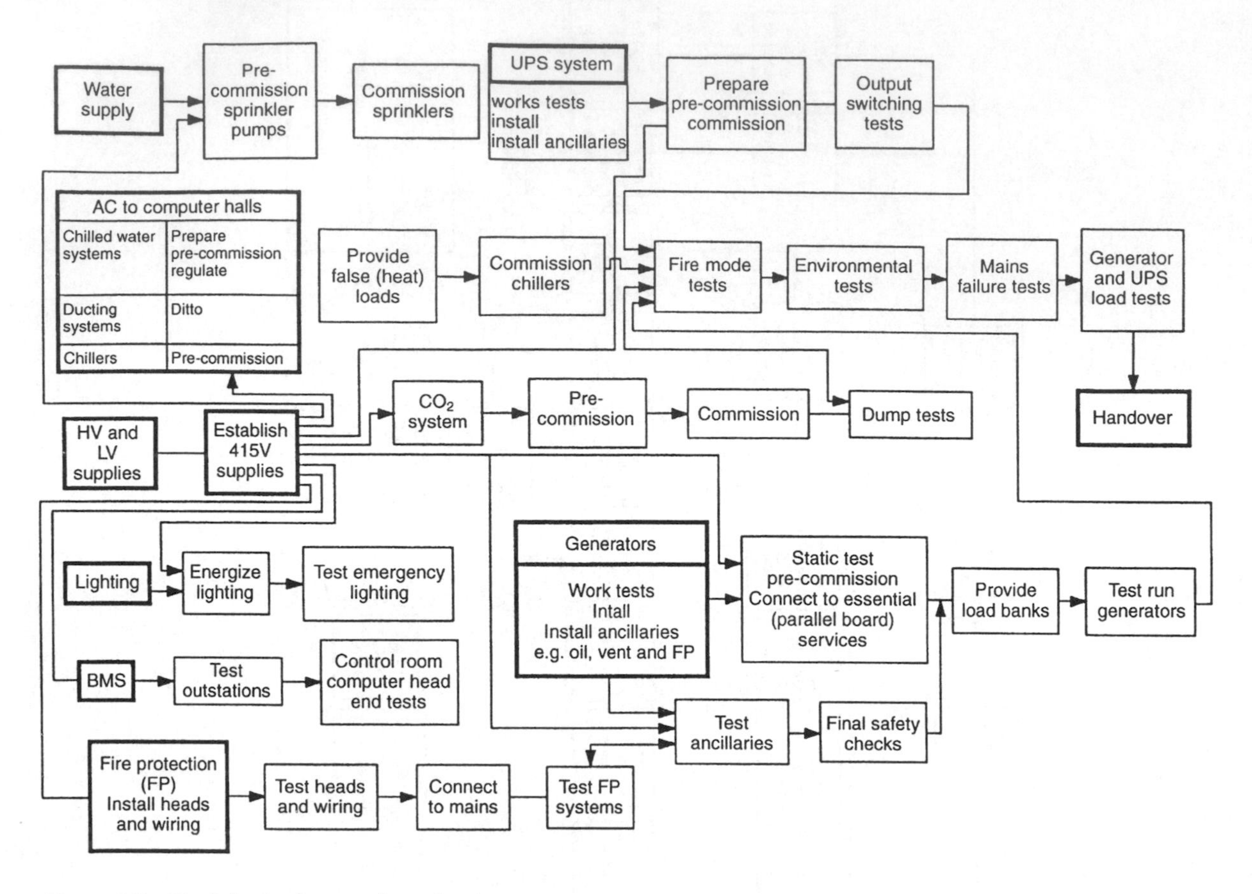

Figure 9.5 Block logic diagram for a bank computer centre.

go into the detail that is now necessary to carry out the work. None the less the contractor may be unable to help. A particular logic may have no flexibility in it, e.g. its technology does not permit doubling up the resources to halve the activity duration.

During negotiations with the BS contractors and specialists to bring a commissioning programme to fruition, the site manager cast in the role of commissioning manager may find that utilities will not be available when required. Gas, water, electricity and drainage are critical to commissioning, with being electricity of paramount importance. Without electrical power to serve the plant rooms and drive fans and pumps out of the motor control centres, systems cannot be prepared for pre-commissioning. To do so using the site electrical systems may overload their capacity and cause interruption and failure of supply. Gas is equally important; the flushing of heating and hot water systems has been proven to be more effective with warm water. Drainage must be connected to facilitate the discharge of products from flushing and chemical cleaning. Any PC who has arbitrarily shown those utilities on his programmes without checking the required timing of their availability with the BS contractors has created an unnecessary difficulty for himself.

Providing gas, water and electricity to the site is but part of the equation. Managing their distribution to the parts of the building where they are needed is the other part that should have been taken care of in construction. With row upon row of lighting fittings being installed and livened up good progress may be apparent, but perhaps to the detriment of running the supplies to the plant room motor control centres (MCCs), something the site manager may only be reminded of by the mechanical services contractor who starts to get concerned that the electrical contractor seems to have his priorities wrong. Space precludes the provision of examples of overall commissioning programmes which on complex projects can run to hundreds of activities. Most will be able to envisage their content from Fig. 9.2–9.5. Meanwhile we will consider Fig. 9.6. It is a variable air volume job of some size showing approximately the last year of the contract period. The client and DT have seen wisdom in not trying to compress acceptance tests (system proving) and any necessary fine tuning into the contract period. The overall commissioning period is broken down into two overlapped groups of activities, the first comprising system preparation and pre-commissioning, the second commissioning in its true sense. The programme indicates that draft and final O and Ms (including BMS graphics) are required six and three months prior to handover. This is essential if they are to be finalized for handover via the PC to the PS on time. Their handover is a legal requirement under the CDM regulations. Progression through the commissioning process is of hands on activities carried out by the installer or his appointed specialist. A similar pattern is repeated for every BS system or group under the control of the assigned commissioning manager, the mantle of responsibility for which may, by default of a proper commissioning management specification, have fallen upon the site manager.

9.4 Making progress

9.4.1 GOING THROUGH THE PROCESS

The specifications for commissioning management sometimes call upon the process to be documented in manual form. The scope and content of a commissioning manual shown in Table 9.4 is typical, but may vary according to the technology of the BS system it relates to. The point of having a commissioning manual is to provide definitive documentation of the process the system was taken through, the results of testing and,

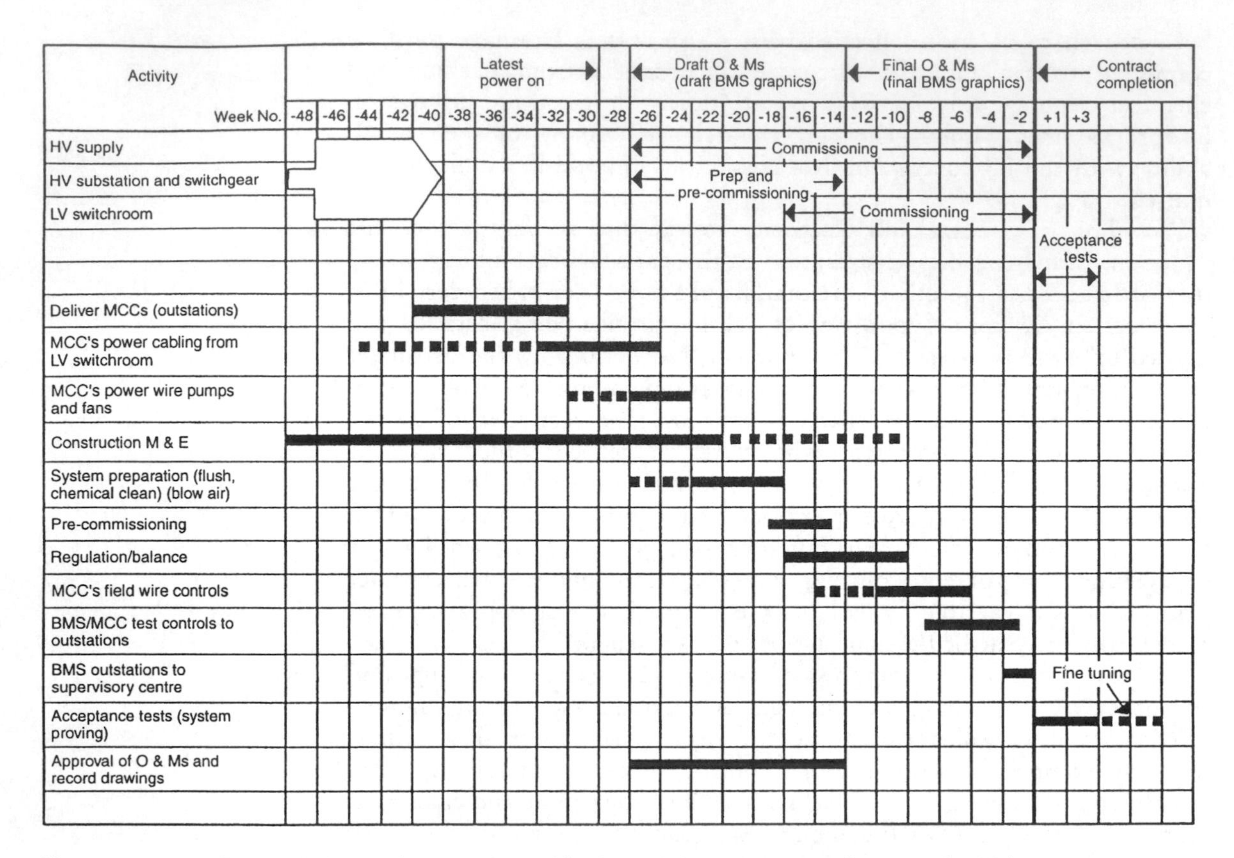

Figure 9.6 HVAC commissioning for a project with six to eight motor control centres (MCCs).

if appropriate, system proving and fine tuning updates. It may also include certificates of compliance for manufactured plant and equipment and any external inspecting authorities. Variants of commissioning manuals may include a record of system defects and their clearance. Manuals of this format may be called upon to be issued as stage reports, building to a final document.

Advice was given in the previous chapter on when to introduce commissioning into the site progress meetings (see Table 8.2). It is through that agenda and the contractor's report, or for those jobs where it is a specified requirement, or the site manager's choice, to hold separate commissioning meetings, that the progression of commissioning will be recorded.

Returning to our vehicle for progression, Fig. 9.6, we will demonstrate the unavoidable logic to the commissioning process. As we go through the process it will be noted that there will be requirements for the

Table 9.4 Typical scope and content of a system commissioning manual

For each system, details comprising but not limited to:

1. Method (including safety arrangements)
2. System schematic
3. System preparation report
   e.g.
   - Cleanliness after flushing and
   - Chemical cleaning
   - Blowing dust out of ducts, wire ways, panels and boards, etc.
4. Water treatment details
5. Pre-commissioning checksheets
6. Commissioning results (initial and final results including details of any changes to systems, e.g. belt and pulley changes)
7. Test results

* 8. Capacity results (e.g. boiler, chiller and generator load tests)

* 9. Environmental testing

** 10. Certificates of compliance (e.g. manufacturers and inspecting authorities such as EMC and NICEIC)

11. System proving
12. Fine tuning

Notes: * These may be covered in 11 and 12.
** If copies only are provided, each manual should state the location of the 'Master', e.g. passed to the PC in the BS Contractor's H & S file.

building spaces to be in a particular condition. We are concentrating on HVAC systems as these appear on the critical path of commissioning most frequently and for longer than any other service. However, comments will be made on other BS systems as appropriate.

All that follows in the subsections of 9.4 apply to HVAC systems that saw their air conditioning elements originate in the USA, fan coil, induction and VAV. Much still applies to the systems now firmly embraced from Europe and Scandinavia, chilled beams and ceilings, and displacement ventilation. The piping and ducting, plant, equipment and terminals have a great deal in common in their commissioning requirements with the air and water systems of America. They differ in their need to be 'proved' and 'fine tuned' under occupation for optimum performance.

The Japanese have taken the development of American unitary systems seemingly out of retaliatory competitive reach with their variable refrigerant volume (VRV) or flow (VRF). Each of the major players – Daikin, Hitachi, Mitsubishi and Toshiba have their own closely guarded variants to the underlying technology. The systems are usually designed by the companies or their approved agencies and installers. PCs and site managers must be aware of the divisions in responsibility these systems create, as they complicate the provision of HVAC through their partial nature. Marketed as air conditioning they are really only comfort cooling systems requiring supplementary fresh air which must be ducted if the building is sealed. The systems present little scope for adding moisture and can give rise to humidity levels nearer the lower end of the comfort range in winter. The ease of commissioning benefits are considerable. They need negligible preparation, limited pre-commissioning and the pipework design, following rigid rules in its layout, is virtually self–regulating. Microprocessor controls to exercise the installations complete the commissioning picture.

9.4.2 SYSTEM PREPARATION

The systems of containment that distribute fluids, gases, including of course air, and wires, also contain the detritus of manufacture and the on-site construction processes. Offsite pre-assembled plant and equipment should have been delivered to the project documented with the quality control procedure through which it had passed, the last stages of which included preparation, final clean and the plugging or covering of open ends of pipes and ducts, etc. However well units are protected in transit, in their passage across site into the building and placement in position, they can be at risk. If the plant and equipment have been delivered early due to the following construction of the building enclosing their access route, they may need to be maintained to prevent deterioration until and beyond the point when they are connected to the distribution systems of which they form the heart. Electrical motors and switchgear must be kept dry and the rotating parts turned over. Even assuming that the best site practices have managed to keep surface water out of the plant room the greatest risk to electrical equipment is through condensation, the ravages of which can, if not prevented, be greatly reduced through the use of chemical desiccants, temporary heating and ventilation. To prevent one risk, another is created. In order to maintain and monitor the condition of early delivered plant and equipment it is necessary at least to partially unwrap it, whereby it becomes exposed to any building works and the by-products of the working environment they create, i.e. dust, dirt and moisture. The risk increases when the services trades commence their activities. To make connections to the plant and equipment further removal of its protection takes place. Now the manufacturer's bungs, plugs, caps and seals are removed to facilitate the connection of pipes, ducts, cables, trunking and conduit and the equipment subjected to the debris of making those connections.

Electrics fare better than mechanical items with respect to the preparation of their services for commissioning. Connections to plant and equipment have more openness to them and can therefore be subject to the progressive inspections of construction to ensure that the burrs, offcuts and cable dressing waste are cleared out. Trunking for wiring is also inspected for internal cleanliness before the lids are put on. Not so with mechanical services. Ducting and pipework that look to be superb examples of the sheetmetal workers' and pipefitters' skill respectively, can hide unquantifiable problems arising from millscale, welding slag, hemp and paste, excess solvent on plastic pipes, and the excesses of sealants and jointing materials on metal and plastic ducting. That is for systems that externally have given confidence in being well constructed. Those systems that have suffered difficulties in achieving the required standard of build quality, necessitating considerable rework following inspection, may hold other terrors. Weld penetration, excessive hemp and paste causing spiders' webs across pipe bores are not uncommon. The uptake of quality systems in the industry has seen a reduction in the numbers of overalls, packets of sandwiches, rubber gloves, tins of solvent, bottles, flange bolts – you name it, we've found it – that have been removed from systems when it was discovered that the media that were intended to flow through them would not pass.

Convinced at least that mechanical systems require some internal preparation let us deal with the easy one first. With power on to the MCCs and field wiring to the fans, they can be switched on to blow air. This is done preferably before the air distribution terminals are fitted. The site manager is advised to ask the BS contractor to inform him when fans are being turned on so that building trades in those areas can be warned. Dust and dirt discharged from ventilation systems have been known to give an unspecified 'stipple' finish where painting is in progress! There can be no doubt that the pipework systems present the greatest challenge in their need for adequate preparation for commissioning. The materials and their related method of jointing in ascending order of preparation difficulty are plastic, copper, iron and steel. Iron, where used in foul and surface water drainage systems, presents less of a problem in its clearance due to the designed accessibility of rodding eyes, manholes and access chambers. Underground externally, it can be subject to damage from site traffic and external works – but that is another story. Steel is the difficult one. The degree of difficulty will vary according to the size of the system and its construction. The good news is the wider use of grooved, clamped and bolted systems pioneered by Victaulic, and tungsten inert gas welding. The latter produces a very clean system compared with the more conventional gas and arc welding methods. Welded joints can generally be applied to pipework above 50 mm diameter and is therefore more likely to be found on the larger systems. For smaller systems, and on larger systems where the pipework is below 50 mm, screwed joints using hemp and paste and other proprietary jointing material prevail. Before entering into the process of preparing pipework systems it is useful to keep in mind why this has become so necessary in recent times. Better insulated buildings, complying with more stringent legislation concerning energy use have meant that for current building sizes and types compared to, those of, say, twenty years ago, pipework sizes and the terminal heat and cooling emitters are much smaller. Coupled with this the flows down narrow waterways are microprocessor controlled by finely engineered valves. The controlled adjustments of these orifices are disturbed by the 'ramping up' of system debris. The removal of this debris is a function therefore of a system's material, its method of jointing, its size (internal surface area to be cleaned by the contained volume passing through it) and the specified process of preparation, by flushing, chemical cleaning and final water treatment. Procedures for the first two activities are covered in the BSRIA Application Guide AG 8/91 [4]. This is commonly referred to in specifications. The author was associated with BSRIA's previous guidance on the subject, Application Guide AG 1/89 [5].

Flushing

The basic requirements for the pipework system status for flushing to commence are:

- The system shall be complete.
- Small bore orifices and automatic control valves, for example fan coil units (FCUs) and air handling units (AHUs) should be protected by 'looping out'. The equipment is bypassed, usually with the aid of temporary flexible connections across the flow and return.
- Containment baskets and fine mesh filters of the 'Y' strainers located throughout the system protecting small bore pipework and control valves mentioned above, are removed.
- The method statement should describe how the system will be divided into sections for the progressive process which must avoid flushing debris from one part of the system to another.
- A permanent water supply must be available.
- The drainage system must be connected.

Chemical cleaning

Used for static and dynamic flushing, water is limited in its ability to clean a system. Where specified chemical cleaning follows the flushing process and may also be a two- or three-stage exercise. The flushing should have been carried out by the installing contractor but it is usual for chemical cleaning to be the work of a specialist. That specialist may also have been charged under the specification to have carried out the flushing, in which case he would have directed the installing contractor to attend upon him in the temporary works alterations to the pipework system and in pump switching.

Chemical cleaning may involve acids or formulated products, polymers and degreasing agents, each targeting different elements of a system's contamination. They are applied sequentially for specified periods varying from 24 to 48 hours. Parts of these durations may be static and parts dynamic running the system pumps. Intermediate steps (chemical change) and final condition may be 'witness' hold points. There is limited scope to play God and to accelerate the process of flushing and chemical cleaning. Increasing velocities, raising temperatures and strengthening chemicals have finite limits within the laws of chemistry and physics which cannot be hurried without losing their benefits. So if the reader considers that a questionable amount of typeface has been devoted to the subject of wet system preparation it is nothing but a reflection of the time it can take in practice. For the site manager this can be a frustrating time. Nothing appears to be happening to the system. There has been a sharp fall off in workforce numbers, the few that are around don't seem to be fully occupied by holding up jars of liquid to the light or dipping things into them. Be patient. The benefits of proper systems preparation are as unquantifiable as the costs expended in chasing muck from one side of the system to another, creating unrepeatable flow rate figures and the need for remedial cleaning; all of which endanger the handover date.

Water treatment

This too is the domain of a specialist. The designer may have pre-selected the process and the firm, or alternatively described it by specification performance requirements. BSRIA's Application Guide AG 3/93 [6] may have been prescribed.

Dependent upon the system to which it is being applied, i.e. open or closed, and the material of its construction, water treatment will be specified to inhibit corrosion, scaling and the fouling of heat transfer surfaces. In open evaporative (cooling tower) systems rejecting heat from air conditioning or process installations the water treatment will have biocide components preventing the colonization of *Legionella pneumophilla*. Pumped closed loop heat recovery circuits, taking heat from discharged air, to preheat fresh air intakes to air handling units or cold water feeds to hot water storage cylinders, are filled with a weak glycol/water mix.

Holding the lead responsibility for safety on his site the site manager must ensure that chemical cleaning and water treatment method statements, including the chlorination of domestic water services, address the hazards associated with the delivery, storage, deployment and disposal of the chemicals. When water treatment is complete it must be maintained until the end of the contract when the responsibility will be passed to the building user through the H & S File, which includes the

O & M manuals. It is quite possible on large jobs with many separate domestic water systems that in order not to lose the benefits of chlorination, tap running and cisten flushing regimes will need to be introduced to avoid the stagnation of water in tanks and pipework systems.

9.4.3 PRE-COMMISSIONING

Pre-commissioning checks that each piece of plant and equipment and the system that it drives or forms some part of, particularly those that are dynamic such as our HVAC example, is finished and ready to be turned on. Pre-commissioning is carrying out a set of specific checks with emphasis on safety as always, first for people and secondly for the property that is being built. A few designers schedule the pre-commissioning checks to be carried out, others ask for the installer to submit a proposed list of checks with his commissioning method statement. Others may cross-refer their specifications to the lists contained in the BSRIA application guides. All designers are concerned with the complete BS system, leaving detailed checks on plant and equipment, e.g. chillers, boilers, generators, switchgear, etc., to the providing manufacturer/supplier. As part of their tender they will have been called upon to pre-commission and then commission what they have provided. Other than these specialists it is recommended that the installer be responsible overall for setting up the system and declaring that it is safe to operate it. In a multi-service contract he will coordinate with electrical departments to ensure that the supplies to the plant and equipment are safe. The commissioning manager should coordinate and collate all of the method statements and pre-commissioning checklists.

With permanent water, gas, electricity and drainage connected or satisfactory alternative temporary arrangements made, plant and equipment can be individually commissioned and properly guarded fans and pumps brought into service so that commissioning proper can commence through the regulation of systems.

9.4.4 REGULATION

Water and air media of HVAC systems must be regulated to create a built environment or serve an industrial process and perform as the designer has determined. This also applies in creating the appropriate flow rates at the draw offs from domestic hot and cold water systems. Other pipework systems of fluids and gases to laboratories are regulated by pipework design, pressure and flow regulators. Sprinklers are a very special case and are perhaps the finite example of hydraulic design of the pipework system to give density of flow over a unit area, obviating the need for regulating devices. Rainwater and soil, waste and vent pipes are also self-regulating. Compared to systems of local geography, e.g. hosereels in service cores, soil, waste and vent pipes (SW and V) in toilets, the environment-creating nature of HVAC systems requires water to be circulated close to its point of use in a radiator, fan coil unit or air handling unit and determines the need for the regulation of stable flow rates. This is a slight oversimplification, but most HVAC system require the regulating devices for air and water systems, be they fixed (manually set) or motorized (automatically controlled), to be in the fully open position at the commencement of regulation. This is how they should have been set to facilitate the full bore system preparation activities. In the case of motorized valves and dampers the linkages would have been disengaged from the drives, and the ports and blades respectively manually stroked to the open position.

The project specification should give the references of the CIBSE commissioning codes [7] and BSRIA application guides [8] to apply. The codes may be considered as the skeleton or framework of what has to be done, while the application guides provide the flesh and sinew of how it is to be achieved.

HVAC systems are usually balanced to a flow rate tolerance of between +10% and –0%, although other criteria may be specified. Upon completion of regulation the designer will want to see that the resultant flow rates presented to him are repeatable. For the system to function as specified for the lifetime of the installation these are the flow rates that it must produce day in and day out. The witnessing of flow rates either fully or on a 'spot check' basis becomes a quality control hold point. Over time the condition of HVAC systems does deteriorate; they are, after all dynamic and subject to 'wear and tear'. Filters will get dirty, increasing system resistance which means reduced volume. Fan performance will fall away, damper linkages wear, control valve efficiency reduce, and the debris of frictional fluid flow in water systems and dirt in ventilation systems all serve to degrade the original recorded performance values. This is why the results of commissioning are important in their inclusion, as a bench mark, in the O & M manuals.

9.4.5 MAJOR PLANT

It will not have gone unnoticed that heat may have been applied to the system during the preparation activities of flushing and chemical cleaning. To do this the boiler plant will have gone through the pre-commissioning and commissioning process as a local set of activities prior to the main thrust of system pre-commissioning and regulation. Other plant and equipment may also have to be specially sequenced in a similar way. It is not unknown for chillers, boilers and switchgear to be called forward for operation under manual control for 'drying out' or in order to provide an agreed internal climate for the application of materials used in the finishing trades of floors, ceilings and wall panelling. Where these are requirements they can add another level of complexity to commissioning management as systems in temporary use are brought into final condition. Commissioning management is not a soft option career.

9.4.6 CONTROLS

Look again at the strategic programme in Fig. 9.1. Not without difficulty we are trying to hold the logic of sequentially overlapping system preparation, pre-commissioning and regulation. Now let us turn our attention to the controls. These were deactivated and isolated during regulation to avoid their acting automatically and affecting the flow of air and water. The regulation of flows are carried out using the installed devices, i.e. commissioning valve sets on pipework, and volume control dampers for ducting. Automatic controls are provided to regulate the flows for that infinite variety of load conditions created by daily and seasonal changes in external climate and changes in occupancy usage. In parallel with system preparation, pre-commissioning and regulation the installation of activators for automatic valves and dampers, the fixing of sensors and the wiring out of all of these from the control sections of MCCs, can take place. Now with regulation complete system by system, linkages to valves and dampers can be reinstated and each control device and sensor set up according to the manufacturer's requirements. System by system the design functions can be verified until all that are served from the same MCC have been proven.

The demonstration of specified control function may be another hold point to be witnessed by or on behalf of the designer. On large jobs with many MCC outstations and systems served therefrom it is preferable to knock these off outstation by outstation, and reduce the load of observations that have to be made via the control room housing the BMS.

9.4.7 BUILDING MANAGEMENT SYSTEMS

Every few years the sophistication of BS system controls goes through a sea-change of advancement. BS system intelligence is becoming more distributed, of higher level capability, particularly in its interactive capacity. For a long while it has been possible to integrate fire, security, energy and building operation controls, providing they were all of the same manufacture. Now we have enabling technology allowing control systems and components within them, of different manufacture, to be capable of talking to each other.

There are other worthy works of reference but the BSRIA body of documents below provides the most independent of insights into the subject of how a BMS should be designed, procured, installed, commissioned and handed over. *BEMS Book* RN1/90 [9] spawned *Standard Specification of BEMS* [10], *Guide to BEMS Centre, Standard Specification* [10] and *Commissioning of BEMS: A Code of Practice* [10].

Regrettably, many BMS specifications are not well written. They vary from the oversimple to the overspecified, the former suffering from lack of designer knowledge and the latter because the designer, without naming the firm, has written the specification around a particular supplier; and only that supplier is capable of meeting their requirements. What purports to be an invitation to open tendering is in fact a disguise for a single tender.

So where is all this taking me? asks the site manager. The answer to that question is if at some point in your career you get deep into the difficulties of delivering the technology of a BMS system and what is being provided is not being accepted by the designer/client, then you use the BSRIA documentation as a guide to evaluate what should have been specified and what your system is capable of. If you are in that level of difficulty you will need help and there are independent consultants who can provide this service, as can BSRIA.

Optimistically for most site managers the situation will not get that bad and we will return to demystify Fig. 9.4, where the activities are divided into four groups:

- system design
- software design
- site installation
- system commissioning.

Many activities overlap and system and software design take place off site, but they are not activities that the site manager should ignore. There are words in the boxes of both those groups which will be familiar to the site manager: 'works test control panel' and 'witness and approve software test', both at the end of their activity groups. These should be focused upon by the site manager who should call upon the BS contractor providing the BMS to programme them as hold points. Obviously this means all proceeding activities also have to be programmed. With these programmes to hand, the site manager can monitor what is happening off site by questioning his contractor. The site installation activities should present no difficulty. MCCs are shown on Figs. 9.1, 9.3 and in our detailed programme in Fig. 9.6. The activities in the last group on Fig. 9.4, system commissioning are those of greatest importance to the commissioning management firm or manager, for they are to be integrated into the backwards looking programme. If there is one key activity in this last group it is that of writing the BMS witness schedule. Only with that schedule to hand can the duration of demonstrations be assessed for all those functions that the client/designer will wish to 'witness' before accepting the system. Without this schedule the whole process of commissioning is impossibly open ended. To stand in the control room while the designer invokes the meaning and intent of his specification clause 'to witness the demonstration of the functions of the system' when you consider yourself a few hours or at worst days away from handover, is not an experience to be recommended. As the designer goes through the 'schedule' that is only in his mind, a small army of people are sent hither and thither into plant rooms, voids and false ceilings turning plant on and off, and deliberately creating false settings on sensors. The betting is a few gremlins will appear, confidence levels will drop to the floor and signals posted that witnessing will take weeks. Schedules must be requested of the designer so that the controls can be demonstrated system by system back to their local outstation (MCC). Finally the schedule should state what signalling is to be demonstrated between the outstation and the head end computer, which should be in a secure location.

9.4.8 FIRE ALARMS

The site manager should try to ascertain, if possible directly from the designer, whether the fire alarm and control system is basic, meeting only the needs of the fire authority, or whether it is being provided at some higher level of sophistication, mainly because the technology is capable of doing more. Beware the systems of added sophistication. Once you have them the fire officer will want to see them proven. If they don't work first or second visit, the limits of the fire officer's patience may be reached with an icy remark, 'I didn't ask for all this wizardry, don't waste my time, don't call me again until it's 110%'. This is not a criticism of fire alarm systems which have done much in their addressability and diagnostic attributes to reduce spurious signals. It is levelled at systems which are not modular but require the writing of special software that in itself is defective or is let down by the installed hardware. It is in the unproven elements of integrated job specific fire control systems where difficulties manifest themselves.

9.4.9 SMOKE PRESSURIZATION

Of the two methods of controlling smoke,

1. extract ventilation, either powered or natural and
2. by pressurization

the second gives the site manager the most cause for concern. The removal of smoke by extract ventilation is replaced by air from any source purposely designed, e.g. the automatic opening of shopping mall doors, low-level inlet louvres, and is supplemented by air movement through any leakage path. When you reverse the process and pressurize a means of escape route, it is the air leakage paths that can be the source of major problems. It is relatively easy for the BS contractor to install a pressurization system and demonstrate that it meets specified requirements of volume and pressure at the ducted point of discharge. The problem is in the shaft in which it has to perform. The pressurized staircase is building work. So the site manager should check the architect's and structural engineer's drawing and if necessary involve the DT to verify that if constructed in accordance with their design the pressurized zone is capable of retaining the designated pressure differential.

If the design is satisfactory that leaves the site manager with the clear responsibility of achieving an adequate standard of workmanship. Designs where it is difficult to achieve the requisite standard are those that have many openings off the treated area. Lift lobbies containing in addition to the doors leading into the operational areas of a building, but which also have doors to toilets, cleaners' cupboards and services riser ducts, call for stringent construction quality control. As pressurized staircases serve for construction access routes for materials and workforce they are often some of the last areas to be finished in a building and the site manager should plan for the worst likely situation; that will be to allow time for remedial work, 'crackage' closure and retesting.

9.4.10 ELECTRICAL LIGHTING AND POWER

The general electrical lighting and low voltage power usually present few problems to the commissioning manager, and can be carried out in parallel with the mechanical services. Priority must be given to testing and enlivening the feeds to the mechanical services systems that are on the critical path of the backwards looking commissioning programme.

9.4.11 IT AND ELECTROMAGNETIC COMPATIBILITY

The constant churn and advancement of technology must be kept in focus by the PC. Do not always assume that HVAC will remain on the critical path to completion. The incorporation of structured voice, vision and IT cabling in copper and fibre, and the certification of electromagnetic compatibility compliance may take the longest. To ensure electromagnetic compatibility optic fibre cable connections need clean, dust free working conditions if repolishing the ends is to be avoided as faults arise in testing. To test each point can take up to five minutes and structured cabling installed as a BS is almost certain to be of hundreds or thousands of points in magnitude. That's a lot of time.

9.5 Bringing it all together

9.5.1 GENERAL

This clause could, alternatively, be headed 'all together now'. As each BS contractor on the job with his own specialist subtraders for commissioning plant and systems reaches approved status, so we enter the stage when the specified interdependencies between systems (read, also contractors) has to be proven. One of the most important skills the commissioning manager must have is to plan and programme so that interrelated systems required to be demonstrated are available, in sequence as close in time as is possible. On Fig. 9.1, with commissioning detail, this is titled 'Specific tests, demos, capacity, environment and noise'. These are but a few words of what acceptance tests may be called up in the specification. A typical example of acceptance test interrelationships is shown in Fig. 9.5.

9.5.2 MODE OPERATION

Further examples are acceptance tests of plant, equipment and systems running in standby and emergency (or back up) mode. These usually apply to part or whole of the project's electrical infrastructure. An example of this modal operation is where on-site electrical generation lops power peaks to keep maximum demand meters from rising to predetermined punitive levels. In the event of mains failure UPS systems would be switched in to run the IT systems. In such an emergency, generators may also come on and serve the essential services portions of the main switchboards. It varies, but essential services are those which will allow a safe and secure but reduced level of continuing operation. By that definition fire, security, emergency lighting and life or product threatened services for hospitals and supermarket frozen food cabinets would be maintained. In commercial developments in addition to their IT systems an amount of low voltage power, say 10% and general lighting, say 25% may be maintained. It is these sorts of overall system operation modes that the client and DT will want to see demonstrated. The lower order of standby arrangements of duplicate fan and pump sets will usually have been tested progressively at each MCC outstation and again possibly in demonstrating the auto changer function as a hierarchical alarm on the head end computer.

9.5.3 CAPACITY TESTS

If these are important they will, in most cases, have been specified as being demonstrated off site at the manufacturer's works, and witnessed before delivery. Into this category fall major items of plant, chillers, air handling units, high voltage switchgear, large air compressors, etc., etc. The capacity of fans and pumps will have been tested under regulation and balanced to provide the total system flow rate. The capacity of electrical feeds serving mechanical systems are in many cases tested by the switching on of those systems and this also applies to the 'livening up' of electrical lighting systems. By its very nature the on-site electrical generation plant, by virtue of its importance in supporting the building in the event of mains failure, is usually subjected to an on-site load test. This is done by connecting it to mobile (lorry mounted) resistive load bank equal to the generator's capacity.

9.5.4 INTERNAL ENVIRONMENT TESTING

Mainframe computer rooms, often remote complexes serving commerce and industry, are regularly specified as having their internal HVAC environment proven before computing equipment is brought in. Electric convector heaters of capacity and location to match the configuration of the computer equipment are placed in the hall and wired up temporarily. The establishment and maintenance of a stable specified environment for a continuous period of up to 72 hours is a test of the HVAC system and its capacity.

Another example on the mechanical services front is the testing and proving of a sprinkler system's flow rate capacity. This is achieved at the control set providing there is a large capacity drain nearby.

It is not only the capacity of BS that are tested prior to handover, but sometimes the capability of the building envelope. Clients have become very aware of the cost of energy that can leak through a building envelope. Construction specifications are becoming more stringent. The external envelopes of large trading and storage halls of air conditioned super- and hypermarkets are called upon to be 'airtight' to limiting leakage rates. For the site manager this all adds time to the programme as the workforce vacates the building, all doors are sealed and a large mobile fan pumps air into the building. These activities, the location of crackage and its sealing bring real meaning to the term 'build right – build tight'.

9.5.5 NOISE

There is noise testing for both the internal environment and the external ambient condition. The environmental health officer may have imposed upon the designer 'not to be exceeded' noise levels at the site boundaries, emanating from the building. Sometimes when running under emergency generator power the EHO will allow the norms to be exceeded for what is expected to be a short duration. Differing allowable levels of noise may apply to daytime and night-time. All of these have to be proven and witnessed by the parties declaring an interest. The tests are usually conducted at night with all the normal daytime operation BS systems running. It is not possible to carry out these tests during the day and get true readings of the impact of the project on the locale against a background of traffic noise and ongoing construction work.

9.6 The paperwork

9.6.1 GENERAL

Commissioning cannot start without the paperwork that resulted in the production of a backwards looking programme, nor approved method statements covering its implementation. From its start, since its progress is not quantified by linear or volumetric measure, the documented evidence of results and status of progress must be recorded. It is a fortunate site manager who is supported by the specified assignment of a commissioning manager, or who has taken the responsibility for appointing one. Less fortunate site managers must keep track of the paperwork and through it monitor progress themselves. The paper is prolific. It continues from the test results of construction, certification with which plant and equipment arrives on site, continues with the results of commissioning proper, ending with the certification by the client, DT and external inspectorate. The site manager will perhaps now begin to appreciate the benefits of a commissioning file or manual for each system, with a separate one for those combined system acceptance tests of capacity, noise and mode operation. All of this should end up with the O & Ms in the BS H & S file handed via the PC to the PS for onward passage to the client or building user, as directed. Most of the above will be of academic interest. What really concerns the site manager is whether all stages are reaching and being released at their programmed hold points. It is critical to the success of the commissioning process that a system of status reporting is set up and monitored at the site progress or commissioning meeting; see

- Table 8.2, Agenda for Site Progress Meeting from Mid-Construction to Hand Over.
- Appendix K, BS contractor's report.
- Section 9.2.3, referring to hold point reports.

9.6.2 STATUS REPORTS – EXAMPLES

Figure 9.7 is an example from Drake and Sculls' QP. Progress against witnessing the hold points can be monitored. Each test sheet or certificate is in itself evidence of status as can be seen from Fig. 9.8, Drake and Scull's test certificate for a gas installation. This sheet by a CORGI registered installer is a combined safety inspection, specification conformity and test sheet. Figure 9.9 shows mechanical commissioning for an office building with three BS contractors, Rosser and Russell (R & R), Balfour Kilpatrick (BK) and for the controls and BMS, Berkeley Environmental Systems (BES). Note the structuring of the sheet. It emphasizes the importance of collecting systems in groups related to the MCC. The work of the mechanical and electrical contractor must be coordinated so that 'power on' to the MCC and the field wiring from it are installed, inspected, tested and witnessed ready for the HVAC systems to be run. It is worth noting how the BMS is also a staged activity.

On the same job Fig. 9.10 is self-explanatory. On jobs with many BS systems this is the only way to go. The continuous searching and researching through sheaves of paper to establish the status of systems does go on, but the time and effort that is spent by the end of an 'unreported' job defies belief.

A final example is Table 9.5 carried out with a multi-service BS contractor. The prison project comprised 70 plant rooms, 64 boilers, 108 water (heating circuits) and 289 air (vent plus a few air conditioning) plants. The figure shows the status of HVAC commissioning for the 17 buildings. This final example shows it is the only way to give accurate percentage predictions and properly measure the commissioning process. Similar status sheets can be produced covering the commissioning progress of any BS system or groups of installations.

At the end of the project the status reports are on the project file and the commissioning results and certificates are in the O & Ms – see Chapter 11.

9.7 Design problems

In section 8.4, under the subheading of design, we dealt with the risk to the project of the services failing to fit and left the issue of performance to this chapter. The records of commissioning status are very important to the BS contractor and the site manager for those instances where the system is unable to achieve the specified criteria of flow rates, internal environment (temperature, humidity, levels of illumination and noise), or capacity and mode operation. The records serve to point up those moments in project time when unforeseen delays strike.

It is possible that despite the BS contractor having installed the systems to the specified requirements and proven through their construction that they fit into the building, some aspects still don't work

INSPECTION & TEST PLAN (ITP)

Project:- TADCASTER SWIMMING POOL									ITP No.ITP/003
Description:- COMMISSIONING			Zone / Area AS APPLICABLE						Page 1 of 3
Item	Activity	Acceptance Criteria	Inspection S-C	D&S	M/C	CLT	Test Required	Verification Document	Notice Required
1	CALIBRATION	BSRIA BS 5750 QP 10.2	H	R	R	R	-	CX 130 CALIBRATION CERTIFICATE	-
2	STATIC - PRE-COMMISSIONING CHECK - AIR	BSRIA BS 5750 QP 10.3	S	S	R	R	VI	CA/1	
2.1	DUCT VOLUME SUMMARY	"	H	H	W	W	T	CA/2 Q.28	
2.2	TERMINAL SUMMARY (GRILLES/DIFFUSERS)	"	H	H	W	W	T	CA/3 Q.28	
2.3	VAV TERMINAL UNIT SCHEDULE	"	H	H	W	W	T	CA/4 Q.28	
2.4	FAN COIL UNIT SUMMARY	"	H	H	W	W	T	CA/5 Q.28	

INSPECTION & TEST PLAN (ITP)

Project:- TADCASTER SWIMMING POOL									ITP No.ITP/003
Description:- COMMISSIONING			Zone / Area AS APPLICABLE						Page 2 of 3
Item	Activity	Acceptance Criteria	Inspection S-C	D&S	M/C	CLT	Test Required	Verification Document	Notice Required
2.5	FAN (AHU) DETAIL & PERFORMANCE (INCLUDING SPLIT & PACK- AGED TYPE AIR CONDITION- ING UNITS)	"	H	H	W	W	T	CA/6 Q.28	
2.6	CV TERMINAL UNIT SCHEDULE	"	H	H	W	W	T	CA/7 Q.28	
2.7	CV TERMINAL UNIT SCHEDULE	"	H	H	W	W	T	CA/7 Q.28	
3.0	SOUND LEVELS (NR)	BSRIA BS 5750 QP.10.3	H	H	W	W	T	CS/1 CS/2 Q.28	
4.0	ADDITIONAL SCHEMATICS	BSRIA BS.5750 QP.10.3	-	-	R	R	-	CU/1	

INSPECTION & TEST PLAN (ITP)

Project:- TADCASTER SWIMMING POOL								ITP No.ITP/003
Description:- COMMISSIONING			Zone / Area AS APPLICABLE					Page 3 of 3
Item	Activity	Acceptance Criteria	Inspection S-C	D&S	M/C	CLT	Test Required	Verification Document	Notice Required
5.0	HEAT LOAD TEST SUMMARY	" BSRIA BS 5750 QP 10.3.	H	H	W	W	T	LT/1	
6.0	REFRIGERATION PLANT	"	H	H	W	W	T	MANUFACTURERS OWN COMMISSIONING DATA SHEETS	
7.0	COMMISSIONING MASTER CHECK LIST	"	H	R	R	R	-	CU/3	

Inspection Type	KEY.	
H - Hold Point	S-C Sub-Contractor	ITP Originator:-.W E DUDLEY.....
W - Invitation to Witness	D&S Drake & Scull	Company:-........D&S...........
R - Review of Documentation	M/C Main Contractor	Date:-...........08/04/1994.....
A - Approve Document	CLT Client	Issue:-..........A.............
S - Surveillance	VI Visual Inspection	D&S Quality Managers Approval:-
	T Test	.... [signature] ....

The Commissioning Engineer shall be responsible for all Commissioning static checks and tests etc.

Figure 9.7 Inspection and test plan, Tadcaster Swimming Pool, by Drake & Scull Engineering Co Ltd.

properly. Disbelief reigns and confidence goes into free fall as, on what has hitherto been a harmonious project, polarized attitudes set in. The installer argues, 'It's all in to specification: the designer sized the cables, pipes and ducts, terminals, plant and equipment. He also sized the pumps and fans. It's not my fault.' The designer finds it hard to believe that the system does not work and sends the installer off with a whole raft of things to investigate. These 'things to be looked at' are rarely issued under an architect's instruction. The job has gone into delay, money is now expended at a frightening rate and may be unrecoverable. Of course, it is possible that the designer is right and some aspects of the installation have failed despite passing through all inspections, tests and checks laid in their path. Regrettably, the history of the industry is charted with failures that have been as much the designer's as the installer's. The worst situation a project can face is where the installer is sent off on a search for solutions that just do not exist in the design, which is incapable of meeting its own specified performance. Guidance on what to do is given in Chapter 12.

Design failure exposed during the commissioning period is a most difficult situation for all and emphasizes the need for those firms that

Drake & Scull
A JWP Company

TEST CERTIFICATE – GAS INSTALLATION

Project:..

Address of premises tested...

....

Date of test :-..........................

BACK BOILER OPENING		ROOM HEATER OPENING		WALL MOUNTED BOILER	
Chimney swept		Chimney swept		Flue fitted to installation	
Cement rendered		Smoke tested		Casing fitted correctly	
Soot plate fitted		Cement rendered		Air vent fitted	
Pipes protected		Dimensions checked			
Flue liner fitted and sealed		Closure plate			
Vent terminal fitted		Fitted and sealed			
Room air vent fitted		Terminal fitted			
Satisfactory ventilation		Satisfactory ventilation		Satisfactory ventilation	

COOKERS		New	Extg.	CONVECTORS	New	Extg.	WATER HEATERS	New	Extg.
Stabilised		INSPECTED		Air vent fitted			Air vent fitted		

New pipe installation plugged and tested to 140 kPa	Complete installation complete gauge tested	Heat output of new appliances checked
Date : / /	Date : / /	Date : / /
By :	By :	By :

Appliances unsafe: ....

Disconnected & plugged: ....

Reason: ....

Reported to: ....

Comments:

Signed.......................... for..................... date..........

Signed.......................... for..................... date..........

Signed.......................... for..................... date..........

Form:- IG/2

Figure 9.8 Test certificate for a gas installation, by Drake & Scull Engineering Co Ltd.

Criterion Piccadilly		COMMISSIONING STATUS REPORT						Date: 12 March 1992	
		R+R		BK	BES	BES	BES		
System	Plant Identification	Mechanical Comm'ning	Wit'sed	Electrical Comm'ning	Sensor Verification	Software Comm'ning	Software Demon'tion	Wit'sed	Comments
MECHANICAL COMMISSIONING									
MCC12001									
Primary Heating	PPH60101/01 & 02	100%	Yes	100%	100%	100%			
Boiler No.1-4		100%	Yes	100%	100%	100%			
Pressurisation Unit		100%	Yes	100%	100%	100%			
MCC12002									
Chilled Water - Primary	PPC70101/01 & 02	100%	Yes	100%	100%	100%			
Secondary AHU Heating	PPH60201/01 & 02	100%	Yes	100%	100%	100%			
Secondary VAV Heating	PPH60301/01 & 02	100%	Yes	100%	100%	100%			
Secondary 7th Floor CHW	PPC70301/01 & 02	100%	Yes	100%	100%	100%			Fecon Unit dp problem
Secondary Office CHW	PPC70201/01 & 02	100%	Yes	100%	100%	100%			
Secondary Retail CHW	PPC70401/01 & 02	100%	Yes	100%	100%	100%			
Secondary Retail Heating	PPH60401/01 & 02	100%	Yes	100%	100%	100%			
Transformer Room Supply	FNS81101	100%	Yes	100%	100%	100%			
Transformer Room Extract	FNE81105 & 81106	100%	Yes	100%	100%	100%			
Basement Plant Room Supply	FNS81201	100%	Yes	100%	100%	100%			
Basement Plant Room Extract	FNS81205 & 81206	100%	Yes	100%	100%	100%			
MCC12003									
Main Office Extract	FNE80505 & 80506	100%	Yes	100%					94% Witness
Staircase Press. West Core	FNS84201/01 & 02	100%	Yes	100%		90%			Duct to be Installed
MCC12004									
7th Floor Fresh Air Supply	FNS80701	100%	Yes	100%					With Auditorium etc
7th Floor Extract	FNS80705	100%	Yes	100%					
Humidifier	HUM80701	100%	Yes	100%		N/A			
MCC12005									
Staircase Press. Atrium East	FNS84101/01 & 02	100%	Yes	100%	100%	90%			

Figure 9.9 Criterion Piccadilly, commissioning status report.

	CRITERION PICCADILLY MECHANICAL SERVICES COMMISSIONING PAPERWORK STATUS REPORT								DATE AND ISSUE: 23/03/1992 ISS. 5
SYSTEM	PLANT REF.	PUMP/FAN MOTOR TEST	CONTROL PANEL TEST	PRE COMMISSION SHEETS	COMMISSION FIGURES	BELT & PULLEY CHANGE	RE-SCAN FIGURES	WITNESS SIGNATURE	COMMENTS
MCC12001									
Primary Heating	PPH60101/01 & 02	Yes	Yes	Yes	Yes	Yes	Yes	Yes	
Boiler No.1-4		N/A	N/A	Yes	Yes				
Pressurisation Unit		N/A	N/A	N/A	Yes				
MCC12002									
Chilled Water - Primary	PPC70101/01 & 02	Yes	Yes	Yes	Yes	Yes			PUMP OVERSIZE
Secondary AHU Heating	PPH60201/01 & 02	Yes	Yes	Yes	Yes	Yes	Yes		
Secondary VAV Heating	PPH60301/01 & 02	No	Yes	Yes	Yes	No		Yes	
Secondary 7th Floor CHW	PPC70301/01 & 02	Yes	Yes	Yes	No				HIGH P.D. IN DOME AREA
Secondary Office CHW	PPC70201/01 & 02	Yes	Yes	Yes	Yes	Yes	Yes	Yes	
Secondary Retail CHW	PPC70401/01 & 02	Yes	Yes	Yes	Yes			Yes	SYSTEM WITH RETAIL CLIENT
Secondary Retail Heating	PPH60401/01 & 02	Yes	Yes	Yes	Yes			Yes	SYSTEM WITH RETAIL CLIENT
Transformer Room Supply	FNS81101	No	Yes	Yes	Yes	No		Yes	
Transformer Room Extract	FNE81105 & 81106	Yes	Yes	Yes	Yes	No		Yes	
Basement Plant Room Supply	FNS81201	Yes	Yes	Yes	Yes			Yes	
Basement Plant Room Extract	FNS81205 & 81206	Yes	Yes	Yes	Yes	No		Yes	
MCC12003									
Main Office Extract	FNE80505 & 80506	Yes	Yes	Yes	Yes	Yes	Yes	Yes	
Staircase Press. West Core	FNS84201/01 & 02	Yes	Yes						GRILLE BALANCE COMPLETE
MCC12004									
7th Floor Fresh Air Supply	FNS80701	Yes		Yes	Yes	Yes		Yes	
7th Floor Extract	FNS80705	Yes	Yes	Yes	Yes	No			
Humidifier	HUM80701	N/A	N/A	N/A					

Figure 9.10 Criterion Piccadilly, mechanical services commissioning paperwork status report.

Table 9.5 Mechanical (HVAC) commissioning status for a prison

	Power			Water Commission					Air Commission			BMS Commission		
Building	LV Power On	Control Panel Power On	Systems Power Available	Flushing Progress	Heat On	Chemical Clean	Water Balance	Water Balance Witness	Grilles Fitted	Air Balance	Air Balance Witness	Controls Comm'sn	Controls Witness	Building
Work Service Unit	100%	100%	100%	100%	100%	100%	100%	100%	100%	100%	100%	100%	100%	Work Service Service
Entry Gatehouse	100%	100%	100%	100%	100%	100%	100%	100%	100%	100%	100%	100%	70%	Entry Gatehouse
Cell Block No 1	100%	100%	100%	100%	100%	100%	100%	60%	100%	100%	100%	65%		Cell Block No 1
Cell Block No 2	100%	100%	100%	100%	100%	100%	100%		100%	100%	70%	25%		Cell Block No 2
Education	100%	100%	100%	100%	100%	100%	90%		100%	100%	20%			Education
Chapel	100%	100%	100%	100%	100%	100%	70%		100%	82%				Chapel
Admin. Block	100%	100%	100%	100%	100%	85%			100%	68%				Admin. Block
Kitchen	100%	100%	100%	100%	100%	60%			100%	50%				Kitchen
Reception and Visits	100%	100%	100%	100%	100%	45%			100%	45%				Reception and Visits
Medical Block	100%	100%	100%	100%	100%				100%					Medical Block
Category 'A' Block	100%	100%	100%	100%	100%				100%					Category 'A' Block
Total Percentage Achieved	100%	100%	100%	100%	100%	72%	51%	24%	100%	68%	35%	26%	15%	

can, to do something about it. The firms that trade in the contractual arrangements of CM and MC, and PMs can ensure that proper technical appraisals are carried out in the design stage to ensure the technology is capable of being delivered. The technical appraisals must be carried out by experienced competent people and will do much to give confidence that the designed technology is deliverable. For the D & B contractor it is naturally all down to them. The PC most exposed is the one in traditional JCT lump sum contracting who may have limited in-house BS resources and be largely reliant on his domestic BS contractors. He must ensure that they bring their best brains to an early technical appraisal.

9.8 Post contract

9.8.1 GENERAL

We are all confident that the building will stand up and that the external envelope will defend it from the elements. We are not so confident that the BS will create and maintain the specified internal environment, or that the protection and detection systems will keep us safe and secure. Our employers may have doubts about the ability of the services systems to allow us to do the work for which we have attended the building. A degree of system proving is always required and some level of reliability established. The purpose of the building, the density of its services and the level of reliability to be demonstrated within the contract period are functions of building type. Within the building there will be a hierarchy of importance. All of this affects the extent to which building services are called upon to be proven and fine tuned within the contract period. The requirements create high risk for the site manager and his ability to meet the handover date.

9.8.2 SYSTEM PROVING

Hospitals, prisons and the computer and dealer rooms of commercial business are building examples with high risk to life, public safety and money for which failure of function has enormous consequences. The easiest contracts to complete are those where the greater extent of system proving is carried out post contract. The benefit of this is that it allows the PC to proceed to the final account stage of the main contract works, reduce site establishment and under a separate, but related, contract work closely with the client, DT and BS contractor on system proving. For other types of buildings, e.g. wafer fabrication plants, hotels and defence establishments it allows the proving to be undertaken concurrently with operational training. All can benefit from these periods. Under operational training the building services are exercised in real use, and the defects of the main contract carried over into the liability period can be mopped up. Those enlightened contracts which create clearly defined post contract periods for system proving should not be abused by the PC or site manager. If they become the carpet under which are swept too many of the main contract defects clients/ DTs teams will become disenchanted and insist upon system proving being carried out under what then becomes a very hard main contract.

Post-contract system proving will be carried out in a period that should be contractually no less tight than the main contract itself. It should have its own specification, logic diagrams and programmes for the constituent parts and an overall programme. There is no reason why the quality, safety and environmental plans of the main contract should not be continued into this post-contract period. Naturally, system proving involves the preparation and approval of method statements in which the building user's staff must also be considered and protected from risk. The precious period of system proving can also be the opportunity for observing that the BS systems in use comply with the CDM Regulations.

Through energy and environmental performance being well established in the minds of the public and the pockets of clients, the specification for testing these aspects may be drawn up from the guidance of BSRIA's Technical Note TN 5/95 [11] and AG 2/94 [12].

9.8.3 FINE TUNING

All BS are exercised by the operation of their systems in the functional operation of the building. Setting aside defects for the moment there is usually a need, in varying degrees, for some fine tuning. This applies most frequently to the HVAC systems, but other examples are adjustments to pressures (and thereby flow rates) in domestic water systems, or in adjustments to the sensitivity of automatic light switching sensors. Having set up systems and got them signed off contractors are reluctant to return to make frequent adjustments for which they are not paid. On larger projects these return visits can become expensive. There is a more harmonious way forward through the supplementary fine tuning contract.

When tendering for the BS works the contractor is called upon to submit a tender that covers for attending a 'fine tuning' site meeting for one day each month for 12 months. That meeting will be chaired by the designer or the client's representative, e.g. facilities manager or works engineer. They discuss the need for any 'fine tuning' based on occupancy feedback. For any adjustments that are different from the specified parameters the BS contractor gets paid according to rates submitted with his supplementary tender. The designer is, of course, paid for his additional input and for the preparation of a report that builds progressively to a final document at the end of the 12-month period.

The benefits in all round satisfaction far outweigh the costs involved and if after, say six months, the environmental systems are satisfactory for say, January to the end of June, then the remaining six months of fine tuning support can be terminated by mutual agreement.

Possibly at the fine tuning meetings the feedback reports will advise of defects which must be remedied by the BS contractor under the defects liability period of the main contract. Should the installing BS contractor also be appointed as the term maintenance contractor with a permanent 'on-site' workforce, the adjustments to be made following a fine tuning meeting may cost nothing at all.

9.8.4 INTEGRATED (LOW ENERGY) BUILDINGS

The future is now and from the analysis in use of pioneering integrated buildings will come the knowledge for new design guides and codes that will make such buildings standard. These 'greener' buildings bring quite different requirements for the commissioning manager to learn. The landmark buildings of

- Gateway 2, Basingstoke
- Queens Building, De Montfort University, Leicester
- Linacre College, Oxford
- Anglia Polytechnic University, Learning Resource Centre
- Birley Health Centre
- The Inland Revenue Office, Nottingham
- Powergen's HQ near Coventry and
- BRE's new low energy office and seminar unit, Watford

have all made new demands on what can be proven inside the contract period. It is, however, in the lessons from their post-contract system proving and fine tuning that they will advance commissioning management. Functions will be tested during the contract, e.g. louvres, light shelves and automatic windows activated. It will take post-contract time for intelligent facades and controls to self-learn a building's thermal properties and occupants' preferences.

References

[1] L.J. Wild for BSRIA (1992) *Design Information Flow*, Technical Note TN 17/92, Bracknell.

[2] L.J. Wild for BSRIA (1988) *Commissioning HVAC System: Division of Responsibilities*, Technical Memorandum TM 1/88, Bracknell.

[3] BSRIA (1994) *The Allocation of Design Responsibilities for Building Engineering Services – a code of conduct to avoid conflict*, Technical Note 8/94, Bracknell.
[4] BSRIA (1991) *Pre-Commission Cleaning of Water Systems*, Application Guide AG 8/91, Bracknell.
[5] BSRIA (1989) *Flushing and Cleaning of Water Systems*, Application Guide AG 1/89, Bracknell.
[6] BSRIA (1993) *Water Treatment for Building Services Systems*, Application Guide AG 3/93, Bracknell.
[7] CIBSE, Commissioning Codes: *A: Air Distribution Systems,* 1971; *B: Boiler Plant*, 1975; *C: Automatic Controls*, 1973; *R: Refrigeration Systems*, 1991; *W: Water Distribution Systems*, 1994.
[8] BSRIA, Application Guides: *The Commissioning of VAV Systems in Buildings, AG 1/91*, 1991; *The Commissioning of Water Systems in Buildings, AG 2/89*, 1992; *The Commissioning of Air Systems in Buildings, AG 3/89*, 1989.
[9] BSRIA (1990) *The BEMS Book, RN 1/90*, Bracknell.
[10] BSRIA, BEMS Centre Publications: *Guide to BEMS Centre, Standard Specification*, AH 1/90 Volume 1, 1990; *Standard Specification of BEMS,* AH 1/90 Version 3.1, Volume I, 1990; *Commissioning of BEMS: A Code of Practice*, AH 2/92, 1992.
[11] BSRIA (1995) *Performance Testing of Buildings*, Technical Note TN 5/95, Bracknell.
[12] BSRIA (1994) *BEMS Performance Testing*, AG 2/94, Bracknell.

Final inspections 10

10.1 Setting the scene

10.1.1 GENERAL

This should be the shortest chapter in the book. Not only the imperfect world but another major change of emphasis in the project cycle determines otherwise. All of the site manager's skills have been stretched by directing his team to deliver the project on time, at the contract price. His good management will naturally have made sure that the specified standards were met along the way. Or were they? Beware, for you are entering a high risk time zone. 'Surely not? Quality control procedures of inspection and witnessing hold points have been met in construction, commissioning is going well and defects are being cleared at a reasonable rate. Why the doom and gloom?' In a word, confusion. The very best site managers recognize the need for a change in emphasis from the general management of defects into the round of final inspections. They control the change and don't let it happen around them. Far too many potentially well managed jobs fail to make a high grade because the site manager has not been aware that client/DT have commenced final inspection without his knowledge. It only needs the BS contractor to have organized the fire officer to witness the fire alarm and the insurance man to witness a sprinkler test and we have management of the project being invisibly taken from the site manager's control. It is essential for the site manager to be proactive and he can only be this through understanding. With the commencement of the first package of construction work defects are recorded in a volume that can easily be controlled. As other elements are brought on line the complexity and total volume of defects starts to grow, but these are still very controllable, being applied to static 'in place' works, such as drainage, foundations, superstructure, lightning protection, roof, floors and cladding. These construction defects can usually be typecast into matters of size, content and appearance. As the number of packages increases so does the number of defects that occur at trade/package interfaces (coordination). By this time in the construction sequence services are appearing on the scene in the plant room areas and down the distribution routes. Finally, with internal partitions, floors and ceiling finishes, services second fix, final connections, pre-commissioning, preparation and regulation, the scope for the total number of defects being dealt with on the site, both in individual packages and at interfaces, is enormous. The early anticipation of the time, and the setting in place of the 'final inspection mode' please, not 'snagging', for recording defects is a critical choice for the site. It is from that moment onwards that the defects lists will contain more deviations of the types related to function, performance and finished product appearance.

The site manager must take the lead in setting up and controlling this change from the 'normal' individual package/trade/work element inspection attitude into the 'final inspection mode'. The site manager must start looking at his job in the same way that the client and DT will look at it. If not, he will be surprised by what they find. They will be taking a holistic view of the completeness of the finished product, not the narrow trade or BS system view that has prevailed hitherto on site.

To be effectively proactive the site manager needs a strategy which, after agreement with the BS contractor, can be rolled out in front of the client and DT. The unfolding of a planned and programmed strategy should give confidence to the client and DT, enabling them to plan their workload for integration into the programme. It will be a great help if everyone can talk the same language in defining defects and their categories.

10.1.2 DEFINITION AND CATEGORIZATION OF DEFECTS

There is certainly confusion in terminology and it would aid the harmony of a project if all parties could agree on what is a defect and for that definition to be used consistently. Definitions may differ but the following seems to pass most tests.

'Defect: An unacceptable level of deviation from specified requirements in one or more aspects of size, content, function, performance and appearance.'

Defects are usually categorized as:

- Inc – incomplete work
- Def – deflect (as definition)
- Dam – damage
- Var – (late variation).

Specific to a project, defects may be further identified by location:

- building
- floor
- area or room.

Completion of the classification would be the addition of the trade or trades involved in its clearance. This does not necessarily mean an assignment of responsibility for remedial cost, but which trade or trades will be involved in bringing the work to standard.

The inclusion of Des: (design) as a category has been avoided as for BS it is considered to be a 'special case' and is dealt with a little later.

10.1.3 THE IMPACT OF CONTRACT TYPE

The MC ManCon 87, does not have a contractual requirement to prepare defects lists or offer the project as a whole for practical completion. There are, however, obligations on the MC to comment on works contractors' notifications of completion and to deal with the architect's instructions to rectify defective work as and when they are received. In essence there seems to be no difference in any of the mainline contracts, construction management, traditional JCT or D & B. This leads to the recommendation that the PC and his site manager should gear themselves up to control the final inspection process rather than drift towards handover.

10.2 Planning and programming

10.2.1 STRATEGY

Contractors have little difficulty with the easily made statements to the effect that their policy is the 'early identification of defects' and that 'these defects will be corrected during the progress of the works and in any event before practical completion'. Doing what they say is the difficult bit. It is easier for the static building works which by their linear and volumetric nature make progress measurement and 'ready for final inspection' status recognized. Yes, here it comes again: BS are different. From the previous chapter we have seen how commissioning goes on to the end of the job and how by nature and or wisdom it is taken beyond. Commissioning involves much toing and froing and playing with the 'finished' article. They never stand still. Systems given their static final inspection weeks ago have been interfered with to demonstrate repeatable flow rates, modal function and performance against false loads simulating climatic seasons other than the current. All of this activity provides scope for the previously inspected work, including the adjacent work of other elements, to be damaged. Late variations also arise from the BS commissioning. Towards the end of commissioning, instruction and training in the operation and maintenance of the BS should be given to the end user's staff and operatives, all to specified details. Here is further opportunity for damage, defect and late variation to undo the work of earlier inspections. Nevertheless if the job is to finish on time final inspections must commence early.

Advice was given in Chapter 8 on when to bring the subject of commissioning on to the site progress meeting agenda and these timings should also serve for starting to organize for final inspection – see section 8.2.3. Too many site managers fall into the trap of not approaching the client and DT at all. They hope they won't carry out any inspections or perhaps because the job is running tight to the handover date, inspections will be a last minute rush around. Those site managers always are, but shouldn't be, surprised when they fail to get their practical completion certificate because the client and DT refuse to be hustled and think the site manager is trying to get away with something – he is. So they take time to inspect the job and create lists of lengths beyond the site manager's worst fears. Resources for defect clearance cannot be mobilized in time. Practical completion is refused and the rumble of liquidated and ascertained damage gets louder. Face up to it; there will be inspections, so get organized. The site manager's strategy should be to talk to the contract administrator with some idea of timing and method, and proposals, for an escorted walk round. If clerks-of-works are present on site, find out whether and when they are to be instructed to carry out the final inspections on behalf of the professional team. It is important to ensure that members of the professional team are escorted by a responsible, project knowledgeable representative from the BS contractors. By his knowledge of the job the BS escort can do much to keep the designer's defects lists to a minimum. A word in the ear about 'last week's variation to change the insulation specification to "hammer clad" in the pump room' and 'don't you remember you gave us a waiver to drop the pipes in that corner' and 'we'll have the belt and pulley change finished by tonight' all serve to keep items off the list that subsequently would have become difficult to remove. The site manager may have to bring pressure to bear on the BS contractor's organization to ensure the proper level of escort is available. But remember the designer is also stretched by involvement in the parallel commissioning and possibly training instruction activities. A key plank in the site manager's strategy should be to get agreement on an acceptance standard.

10.2.2 ACCEPTANCE STANDARD

Historically there has been considerable dispute on the part of clients and their DTs on the definition of what constitutes practical completion. Views range from the extreme of completion with 'nil defects' to the more relaxed position of accepting that the proposed occupier is able to make safe and reasonable use of the building in its incomplete or only cosmetically defective state. In practice this view is determined more by the availability or not of tenant or occupier than by the legal interpretation of the contract. Commercial and retail projects attempted to be handed over at the end of a contract period that has seen the client's markets slump may find themselves being 'nit picked to death'. Clients are reluctant to take over a building for which they have no tenants – involving themselves in costs for insurances, security, heating, running the plant on tick-over, and business rate council tax, etc., etc.

Agreement on project condition at handover should focus on sensitive areas which may need special treatment, e.g. 'a clean room area' which will be in use on a 24-hour basis will need careful attention since re-entry will be restricted. Similar situations will apply to prisons and hospitals. This is not to say that other areas in the building may be less sensitive, and any agreed relaxation should not become an excuse for poor supervision. The agreement will take into consideration the types of defects and their volume. High on the list of priorities would be all items which could affect the issue of an occupation certificate. Among these would be:

- fire protection and alarm systems
- emergency lighting systems
- means of escape – smoke management

The level of acceptance negotiable can have a very real effect on the programme. For this reason it is important to establish this agreement at the earliest opportunity. Figure 10.1 was negotiated by the author for a major new prison.

10.2.3 RESOURCE MOBILIZATION

If the PC's team has been effective and made sure that BS contractors have set up and maintained proper supervisory and inspection levels throughout construction the change to a higher gear will now be smoother. The site manager may not have inspected that arbitrary 10% of the BS contractor's work earlier. He will need to intensify his inspections now, concentrating particularly on priority areas. This means he must have the resources. If the resources can be made available from the team on site this is fine as they will have the all important job knowledge which will help them get through a greater volume of inspection than if new teams have to be drafted in.

A similar situation will exist within the BS contractor's organization. Many of them do not realize that it is the period during which they are most likely to fail. Site office managements are blind to the demands of paralleled working on the key streams of activity:

- commissioning;
- preparation of O and Ms and record drawings;
- training and instruction of end user's staff and operatives;
- final inspection and defect clearance programme.

SHEET 1 OF 2
DATE. 09.10.90

SUBJECT: REQUIRED STATUS OF COMMISSIONING AT BUILDING HANDOVER

To: PBA, PSA, MH, D&C and WCM

1.0 OBJECTIVE
To define the requirements for the status of engineering services commissioning at building handover.

2.0 BROAD DEFINITION
The general requirements is for each building handed over to be capable of functioning safely as a stand alone entity served with power, water, gas and drainage from the site utilities.

3.0 GENERAL COMMISSIONING/TESTING FOR M, E, PHE & FP
3.1 All installation work to be complete and equipment/fittings identified.
3.2 Commissioning – systems and equipment moved from static to dynamic status and functionally tested. Plant and equipment performance results recorded and witnessed.
3.3 Testing (HVAC) – All air (inc. Smoke vent) and water (inc. HWS) systems balanced, results recorded, witnessed and any re work due to fault or design change (such as belt/pulley change) carried out.
3.4 Testing (HVAC Controls) – All controls and BMS functionally tested as specified insofar as weather permits at time of that test.
3.5 Testing (PHE) – All RW, SW & VP's and branches to be pressure tested, recorded and witnessed. All MWS and DSCWS systems and equipment to be pressure tested recorded and witnessed.
3.6 Testing (E) – All HV & LV Switchgear and boards, sub main cables, distribution boards, lighting and small power,

DISTRIBUTION:-

R Casey	TBA	FBS	WCM
G Mulvihill	JJD	JJD	WCM
A Ladd	MH	DWF/PEJ	
G Edler	PSA	RSG/BGM	WCM
A Ames	PSA		

FROM: L J Wild **ORGANIZATION:**

Figure 10.1 Required status of commissioning at handover, Woolwich Urban Prison.

It is, or should be, a period of high adrenalin flow for the achievement of a successful project is in sight. Unless the site manager has started to plan the final inspection and defect clearance programme early enough with the BS contractor he will be making demands too late. Correct plant and equipment power, emergency lighting and alarm systems to be earth bonded, IR and ELI tested recorded and witnessed. Lightning protection.

3.7 Testing (E – Specialist Systems) – Radio, PA. TV; CCTV Telephone, geophone, multiplex systems to be earth bonded, tested, recorded and witnessed insofar as they can function in stand alone mode within the building in which they are installed.

4.0 PERFORMANCE, TESTING FOR M, E; PHE & FP

4.1 Piped Fire Systems – Hosereels, Dry fire risers, sprinklers, fire suppression systems, functional and performance tested to specified requirements and where applicable to insurance and Home Office Fire Officers approval.

4.2 Domestic Water Systems – Hot and cold water systems to be performance tested to agreed method, witnessed and recorded.

4.3 Electrical Systems – Emergency lighting, lightning protection.

5.0 EXCEPTIONS

The following tests have dependencies that exclude them from handover requirements. These tests are generally of a special performance nature and will be carried out when factors such as site activities, operatives training and climatic conditions allow.

- Noise
- Environmental Conditions
- Heat Reclaim
- Heater Battery
- Chlorination
- Catering equipment (eg. dough ovens).

6.0 END NOTE

The transferance and verification of signals from the stand alone BMS and multiplex systems in each building through to the WSU, and ECR room of the EGH will continue past building handover. This will also apply to the specialist systems such as CCTV, geophone, radios, PA and TV and telephone.

LJW/CT/26825
09.10.90

Figure 10.1 *Continued*

though those demands may be the notice will be too short, acrimony will give rise to confusion and fall off in impetus. Just at the moment when success is in sight the site manager's firm grip right to the end will pull him up short of the winning post.

At an early meeting with the BS contractor the site manager should ask how they usually go about mobilizing final inspection and defect clearance. If such a request is greeted by the glazed mid-distance stare from a contractor who has always plodded forward, expecting hassle at the end of the job without really understanding why; then the site manager will at least consider himself fortunate in being able to start the proper management education process early enough. The more enlightened, better managed BS contractor will be able to respond positively saying for example, 'We bring in a special team of electrical inspectors. It so happens I've arranged for them to come in three weeks time. We find this is essential, particularly with differences of opinion we experience over earthing and bonding.' Or the site manager may find that the BS contractor wants to make arrangements for members of his team to accompany the clerk-of-works on his final inspections.

It is a time for the site manager to be tough. The offsite BS organizations can need quite a bit of shaking of their trees for the resources to drop. Quite often their resistance is born out of the awful dawning that an overbilled, profitable job is about to be turned into one of breakeven or loss by the site manager's demands. It is a fault of the construction industry that they cannot understand the considerable amount of money that needs to be expended through resources on the four stream activities above (see p. 237) which appear to do so little to change the constructed work. After all, three of the four activity streams, commissioning, documentation and instruction, do nothing much to change the project's appearance. Nevertheless meeting the specified requirements of these activities must be of no less importance to the site manager and BS contractor than the pulling in of a cable or fixing the toilet seats. They are ignored because they don't seem to be doing anything. To change the industry's success rate all parties from client to specialist subtrader must improve their understanding of how the BS are finished, and plan for their realization.

Contracts that have accelerated to recover time lost in earlier activities, possibly unrelated to BS, nearly always end up in difficulty during the four stream activities instanced above. Whether acceleration has gone well or not there is enforced compression of activities where none in reality is possible. In pricing the cost of acceleration all contractors should consider and include for the knock-on effect of speeding up the final inspection and defects clearance through mobilizing additional resources. The sensible justification of these consequential costs of tail-end mobilization may bring client and DTs to more realistic appraisals when acceleration is due to a variation. The PC will also be presented with more accurate figures for acceleration due to the consequence of his earlier action or inaction.

10.3 Inspection and defect clearance

10.3.1 KEEPING CONTROL

As part of the planning and programming the site manager and BS contractor will have agreed when and where, that is in what order, their own final inspections will take place. On the smaller job and those with large open spaces access for inspection need not be limited. On multi-building sites, those with small cellular spaces or of special designation, e.g. clean rooms, operating theatres, board rooms, laboratories, etc. an order of access may need to be established. The site manager is also advised to discuss the BS contractor's own inspection findings and needs for remedial work so that it can be planned with other trades' activities.

Damage

The site manager must share the blame when confronted by an irate BS site manager who has just discovered on final inspection, damaged work in spaces he completed and left long ago. Too many site managers hide behind the contractual and subcontract clauses of making trades responsible for the protection of their own works. The difficulty lies in the impossibility of being responsible for work in an area inspected and defect cleared weeks ago and in which you are not working. In the greater number of instances any protection attached to the BS work, e.g. on perimeter fan coil units, radiators or convectors, has to be removed at some time for the following trades to do their work. The avoidance of conflict is the more equable distribution of responsibility through the site manager accepting that he must demand of the building trades the same levels of supervisory control that we are saying he should have from the BS contractors. Unless the PC and site manager adopt a site-wide equitable level of control the dustbin of contra charges will overflow with dayworks signed 'for record purposes only'. It is amazing the amount of damage 'nobody is responsible for'.

Keeping control of final inspection and defect clearance means controlling access to the iterative process of inspection, clearance, re-inspection re-clearance. Figure 10.2 may assist the SM to produce a similar chart for his project.

The PC's site management team will not, and should not, inspect all of the BS contractor's work. Concentration on key areas should give an indication as to what the standard will be in less onerous areas. It is recommended that inspection be concentrated on such areas as process services, control rooms, clean rooms, kitchens, toilet areas and public spaces, such as shopping malls, dining rooms, conference suites, etc., etc. Coupled with those pinpointed earlier in connection with safety and security they will all move progress towards acceptability in the client's and DT's eyes.

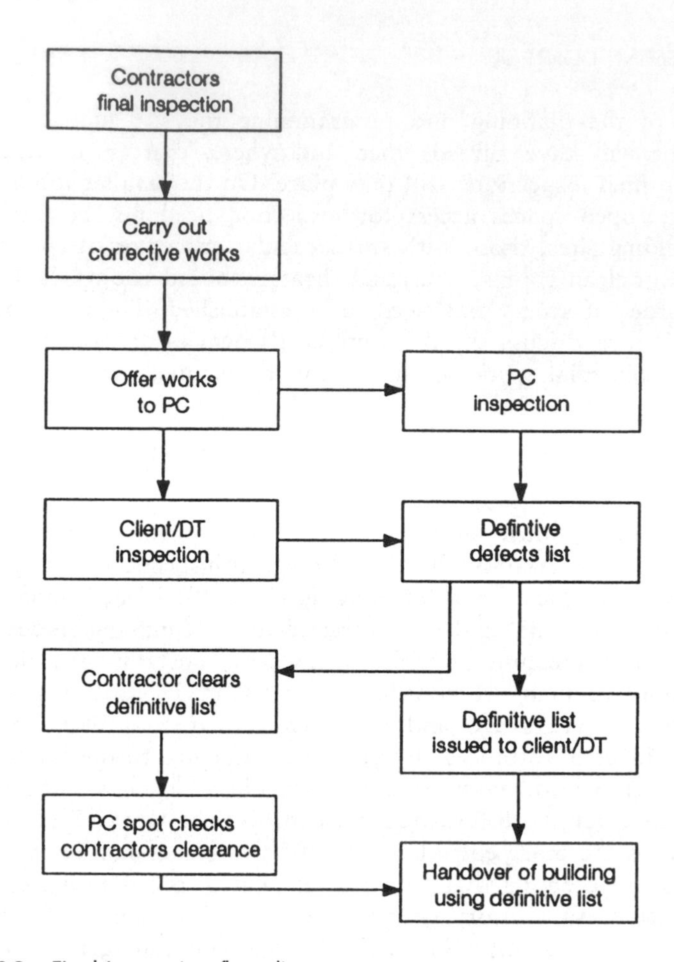

Figure 10.2 Final inspection flow diagram.

10.3.2 KEEPING THE SCORE

There will be many score cards covering the play of the same game. The site manager is responsible for their collection, collation and redistribution to the contractors, for clearance. By whatever process, manually written, typed up or keyed into a computer, the paper work is voluminous.

The computerization of defect records can provide useful flexibility in that reports can be produced by trade, contract and location. The latter can be provided in detail varying from room by room, floor by floor and building by building. The day will come when every site will apply the benefits of computer technology. Without a doubt the ability to present updated information showing rapidly tumbling numbers of defects, has psychological benefits for all.

The fixed to, threaded through and placed on characteristics of BS come into play once more and the site manager may find his services contractors and clerks-of-works are inspecting the works outwards from plant rooms, down distribution routes (if unenclosed and not subject to earlier final inspections) to the terminals. There is nothing wrong with this approach but to avoid overlapping of precious inspection resources the site manager could offer to carry out and record with his resources, those defects to be noted in connection with services terminals – after all the site manager's inspectors will be looking specifically at walls, floors and ceilings and what is fixed on or into their surfaces.

10.4 At practical completion

10.4.1 WHEN AND WHAT IS PRACTICAL COMPLETION?

The site manager would do well to ignore the differences between one contractual format and another. He could argue that practical completion of works packages certified in advance of the project completion as in the case of management contracts should be excluded from client/DT final inspection. On long projects retention on such early works may have been released. The difficulty for the site manager is that for most contracts it remains within the power of the architect, employer's representative or contract administrator to issue instructions at any time requiring rectification of any defect, shrinkage or fault due to the works not being in accordance with contract conditions. So earlier work completed by the lightning protection specialist which subsequently is shown to have a coverplate damaged, probably by a dumper truck that no one saw, remains the responsibility of the site manager to 'sort out'. In a similar way it is usual for the architect, employer's representative or contract administrator to determine 'when in his opinion' practical completion has been achieved.

By now the site manager will understand, particularly if his project is a large one, that there is unlikely to be one single final inspection, but a round of them undertaken in accordance with the strategy planned and programmed, see section 10.2.1. The inspections will be measured against the agreed acceptance standard – see section 10.2.2.

10.4.2 THE CLIENT/END USER WALK AROUND

Unavoidable, but risky for the site manager. Probably having received a flurry of even later variations which have yet to be distributed to the trades the site manager should prepare himself for another wave of instructions following the client's walk round. Here we are not talking about those projects that have been fortunate enough to have a client's representative continuously involved. No, this is the walk round by the client or end user who has at best paid frequent visits, and who is now faced with having to accept the project. These visits give rise to the 'wish list'. Whether they arise from fresh eyes seeing what others have passed by daily, or arise through disappointment at the final product, or pleasure at seeing the potential benefits of adding value, they say, 'I wish I had this, I wish I had that, I don't want that and why is that there?' Providing nothing on the 'wish list' is there as being the fault of the DT, the PC or the BS contractor in failing to meet the employer's requirements so will the approach to the 'wish list' be affected. The execution of items on the wish list will depend on whether they are the fault of the DT, the PC or the BS contractor. The content of the list will be evaluated by the DT. The DT has probably organized the client walk round between the issue of the defects on their final inspections and their clearance leading to the issue of the practical completion certificate. Having ensured that they are not in default the DT may issue all or part of the 'wish list' as variation orders to the PC. In these circumstances the PC may need to search the contract to see if it is possible to refuse to accept the instructions. Alternatively he may pick off those that he can do with the resources on site and recommend that others are dealt with post contract. With sense prevailing a compromise may be reached to the satisfaction of all parties. The site manager must be aware of the difficulties that a client's wish list can harbour for the BS contractor. Additional work can give rise to serious differences of opinion between the site manager and the BS contractor, the former seeing the work as being easy to include, the BS contractor listing a number of difficulties. These may or may not be reasonable. Regrettably it is unavoidable that the site manager sometimes has to go back to the contract administrator and inform him that what appears to be simple wish list requests attract punitive preliminaries costs and could jeopardize handover.

10.5 The defects liability period

In the end the job was handed over – see Chapter 11 – and we are now in the defects liability period. This period is also subject to inspections. The most commonly used contracts have clauses that state something along the line that, 'Within fourteen days of the end of the DLP the Architect or Contract Administrator should issue to the Contractor a list of outstanding defects.' Thereafter the PC has the responsibility through the BS contractor to secure the making good of these defects within a reasonable time. When, in the opinion of the architect, the defects have been made good, he is required to issue the Certificate of Making Good of Defects. The receipt of this usually triggers the release of outstanding retention.

For the BS contractor there are two difficulties with the defects liability period, one of which he shares with the PC. The first is in the dynamic nature of BS. Even passive protective and detection installations are prone to the creation of spurious alarm signals that must be investigated and corrected. They and all moving components in terminals and the larger plant and equipment are prone to historic patterns of failure of which the classic 'bathtub curve' is but one. That curve shows an early pattern of failure, followed by a long period of reliability until, through normal wear and tear, failure again increases. Failures at the front end must be addressed and cannot be left to be dealt with as defects at the end of the liability period.

The second problem, the one shared with the PC, concerns the dispersal of staff and operatives with project knowledge. It may not be within the site manager's powers to stay with the project long after it has been completed. If it has been successful and he has enhanced his reputation he will be off on another job. The problem is therefore one for the firm rather than the individual to resolve. Projects that set up separate post-contract fine tuning BS agreements will derive most confidence from the retention of knowledgeable resources, which by the end of the defects liability period will have seen a seamless change over from construction into trouble free operation and maintenance.

10.6 Design – the special case

There is a major unwritten, unspoken difference of attitude in the expectation of architectural, structural and BS design. The first two appear to be asking for no more than they have specified to be delivered by the PC. The BS designer wants, not only what he has specified to be provided, but also, if it doesn't work, to invoke the 'standard' wording of the specification: 'the installer shall provide everything necessary'. By the dictionary definition of defect – 'a lack of something necessary for completeness; deficiency', the BS designer has specified that the project shall have 'nil' defects, not only in what he has specified but by the addition of anything that he has not specified, but subsequently found to be essential. The inequality in expectation is born of the BS designer's cherished view of his professional status, that too frequently treats the installer as of a lesser order, and yet, when things go wrong expects him to have an equal or higher level of design knowledge, in order to make the job work. It is a situation where the site manager can do much to support his BS contractors and redress the traditional unfairness.

The raising of design as a defect category needs to be handled diplomatically. Who is to say, and has the knowledgeable right to do so from the contracting side, that design is defective? Here at the end of the project we have dealt with matters of fit and tolerance – see Chapter 7. We have dealt with systems that cannot achieve flow rate, function or performance during commissioning – see Chapter 9. Or, perhaps we are still dealing with those problems. Therefore with respect to BS it is recommended that design defects are kept off of the final inspection sheets and are dealt with as special cases. Taking this approach, and it is not unreasonable for all parties to agree to it, enables the BS contractor and PC to define clearly the extent to which they have met specified requirements. Yes, we know we have not been able to show that the HVAC systems, or lighting create the specified environment, but those criteria were not selected by the contractor. It was the designer who implied in the specification the words, 'If you put in these systems comprising the plant, equipment, terminals and distribution capacities and sizes that I have specified – then these environments will be created.'

The subject of design responsibility is a most contentious one which is beyond the treatise of this book. There is much good work of recent times allied to the coming of the Latham Report that bodes well for a more harmonious and less litigious future.

Handover 11

11.1 General

11.1.1 PROBLEMS

Preparation must take place early in the second half of a contractual programme if the historical difficulties of handing over a project, listed below, are to be avoided:

- specified requirements vary from vague to demanding;
- production of documentation is left too late;
- the cost of meeting the specified requirements is usually underpriced;
- it is a contractually onerous process;
- it is a long process with many interdependencies.

11.1.2 THE DISPERSED REQUIREMENTS

As it was with commissioning and its management so it is with handover, the requirements for which will be found dispersed in contract conditions, preliminaries and preambles, and specification clauses. Another trawl is recommended to be directed by the site manager. The BS contractor should look through all documentation in his possession and the site manager should search any other contractual documents which may not have formed part of the BS order. Table 11.1 lists document clause headings where relevant information may be found. Only with that information to hand can its delivery be organized. The main requirements for handover are:

- documentation
- instruction and training
- keys, tools, and spares, etc.

Of these the provision of documentation is the area that usually causes most difficulty. This is broken down into the subelements of:

- operating and maintenance manuals;
- planned preventative maintenance systems (PPM);
- record drawings;
- charts, diagrams and notices.

Table 11.1 The dispersed requirements for handover – clauses where handover requirements may be found

Contract conditions, preliminaries, preambles and specification clauses:
Practical completion
Handover
Commissioning
Operation of permanent systems
Replacement of disposables
Fuel for testing
O & M manuals
Planned preventative maintenance
Technical author
Documentation
Record drawings
As-built drawings
Wall charts
Notices
Parts list
Spares list
Training } Client's/tenants/operators' staff
Instruction } Client's/tenants/operators' staff
Look into *EACH* section of BS specification

Early drafts sometimes graced with the title *Temporary Manuals* may be required to be available for commissioning.

All of the requirements should be met by programming backwards from the handover date; Fig. 11.1 is an example that will be examined in later sections.

11.1.3 HANDOVER AND THE H & S FILE

The HSW [1] provided legal leverage, rarely applied contractually, to ensure that the end user was provided with the wherewithal to operate and maintain the BS handed over to them. Now, with the coming of the CDM Regulations [2] the leverage can be applied more clearly. Unless approved O & Ms and record drawings pass up the chain from BS contractor to PC and on to the PS for issue to the client/end user, the PS will be unable to discharge his responsibilities. This is clearly set out in Clause 125, Stage 5, 'Commissioning and handover' of the *Guide to Managing Health and Safety in Construction* [3]. This states 'Information which is required for the Health & Safety File needs to be forwarded to the Planning Supervisor. This usually includes operational

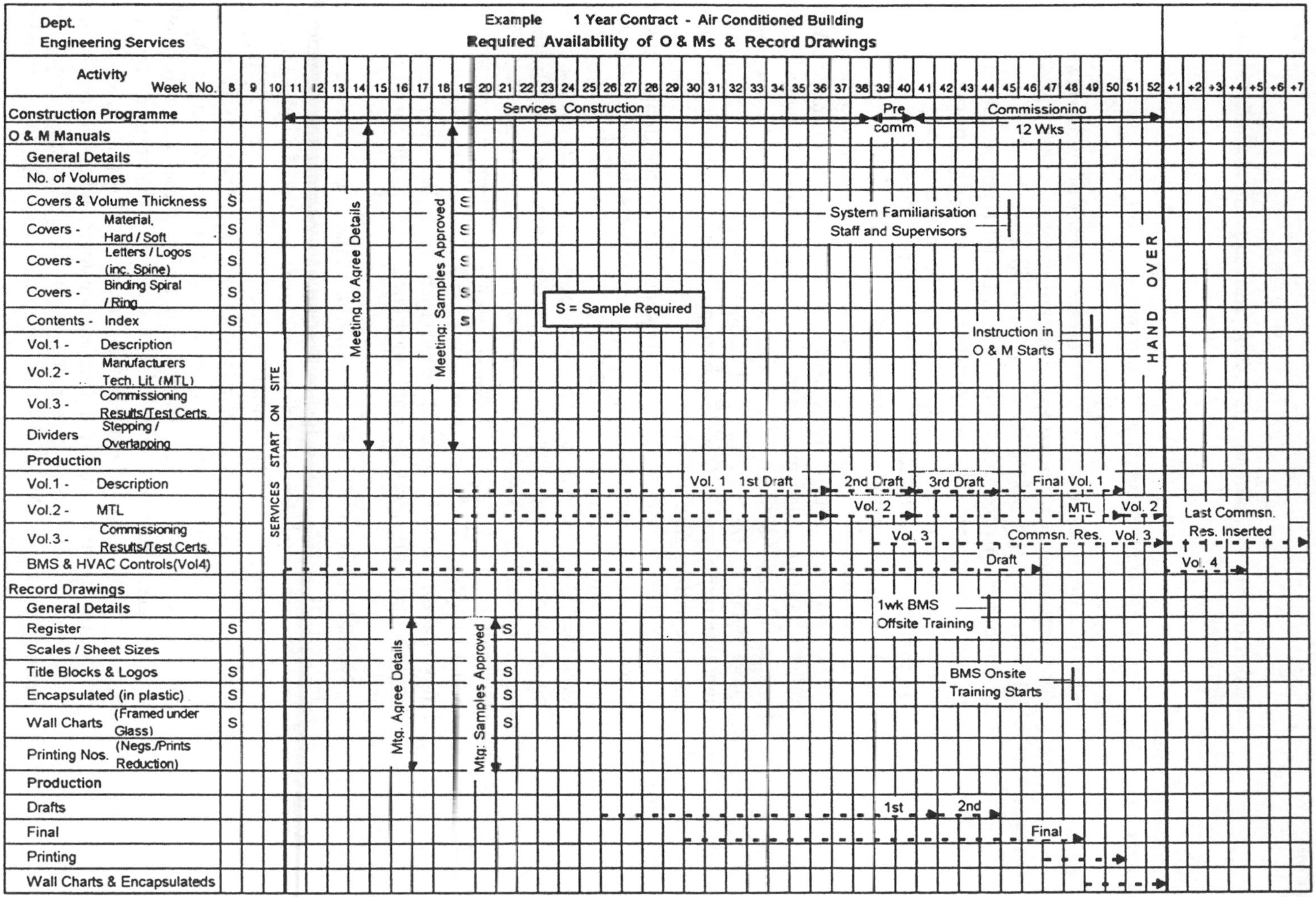

Dept.
Engineering Services

Example 1 Year Contract - Air Conditioned Building

Required Availability of O & Ms & Record Drawings

Activity	Week No. 8	19	21
Construction Programme			
O & M Manuals			
General Details			
No. of Volumes			
Covers & Volume Thickness	S	S	
Covers - Material, Hard / Soft	S	S	
Covers - Letters / Logos (inc. Spine)	S	S	
Covers - Binding Spiral / Ring	S	S	
Contents - Index	S	S	
Vol.1 - Description			
Vol.2 - Manufacturers Tech. Lit. (MTL)			
Vol.3 - Commissioning Results/Test Certs.			
Dividers Stepping / Overlapping			
Production			
Vol.1 - Description			
Vol.2 - MTL			
Vol.3 - Commissioning Results/Test Certs.			
BMS & HVAC Controls(Vol4)			
Record Drawings			
General Details			
Register	S		S
Scales / Sheet Sizes			
Title Blocks & Logos	S		S
Encapsulated (in plastic)	S		S
Wall Charts (Framed under Glass)	S		S
Printing Nos. (Negs./Prints Reduction)			
Production			
Drafts			
Final			
Printing			
Wall Charts & Encapsulateds			

Week No.: 8, 9, 10, 11, 12, 13, 14, 15, 16, 17, 18, 19, 20, 21, 22, 23, 24, 25, 26, 27, 28, 29, 30, 31, 32, 33, 34, 35, 36, 37, 38, 39, 40, 41, 42, 43, 44, 45, 46, 47, 48, 49, 50, 51, 52, +1, +2, +3, +4, +5, +6, +7

Services Construction — Pre comm — Commissioning 12 Wks

SERVICES START ON SITE

Meeting to Agree Details

Meeting: Samples Approved

S = Sample Required

System Familiarisation Staff and Supervisors

Instruction in O & M Starts

HAND OVER

Vol. 1 1st Draft — 2nd Draft — 3rd Draft — Final Vol. 1

Vol. 2 — MTL — Vol. 2

Vol. 3 — Commsn. Res. Vol. 3

Last Commsn. Res. Inserted

Draft — Vol. 4

1wk BMS Offsite Training

BMS Onsite Training Starts

Mtg. Agree Details

Mtg: Samples Approved

1st — 2nd

Final

Figure 11.1 Required availability of O & Ms and record drawings.

and maintenance manuals for plant and equipment, and "as installed" drawings.'

11.2 Operation and maintenance manuals

11.2.1 UNDERSTAND WHAT IS SPECIFIED

The BS contractor must have the greatest depth of understanding of the documentation to be provided. The site manager must discuss with the BS contractor any risk to its timely provision.

Difficulties in providing O & M manuals are not always down to the BS contractor. Far too many projects have inappropriate and poorly written requirements. Sometimes the requirements are vague, leading the BS contractor to produce what he sees as compliant only to be rebutted by the designer saying, 'This isn't what I want', followed by a description of current perceived needs. The BS contractor responds by saying, 'What you have just described is not what is written in the specification. What I have provided meets specified requirements.' Acrimonious debate ensues. If either side gives way they are held responsible for any subsequent delay in the provision of proper documentation. Meanwhile the site manager's site becomes a battlefield and the handover date is at risk. How does this occur?

Too many designers will have concentrated on the design and specification of the systems, turning their attention to the commissioning and documentation needs far too late. Hurriedly they will either (1) simply state that O & M manuals are to be provided, or (2) cobble together a poorly conceived set of clauses, plagiarized from the last job (that itself was produced in exactly the same way). Between these two extremes, that rate on a scale from 'dreadful' to 'awful', there will be found other bad examples. If the BS contractor and site manager have a shared understanding of the O & M requirements that has been acquired early enough they can query its appropriateness and avoid conflict at project end.

Not all of the problems of specifying requirements emanate from the designer. It is well worth trying to establish whether the designer asked or was informed by the client/end user of the latter's strategy for operation, and particularly maintenance. Was it intended that the end user would be deploying in house labour resources to carry out the general maintenance, supported perhaps by term contractors for specialist equipment, e.g. high voltage switchgear, diesel generators, chillers and boilers, etc.? Or will all services be maintained by a term contractor? Writing manuals for the former strategy requires knowledge of the maintenance resource competence level, whereas for the latter a high level of competency can be assumed and more simply written manuals provided. Most term contractors have their own schedules of periodic maintenance for all of the various systems, plant and equipment. If the designers have adopted the HVCA five-volume *Standard Maintenance Specifications* for building services [4], produced in collaboration with CIBSE and BRECSU, then the difficulty of preparing detailed maintenance manuals specific to each project is greatly lessened. These excellent standard maintenance specifications aim 'to provide a benchmark against which to measure the performance of contractors or in-house maintenance departments'. They cover:

- Vol. I: *Heating and Pipework Systems*, 1990
- Vol. II: *Ventilation and Air Conditioning*, 1992
- Vol. III: *Control, Energy and Building, Management Systems*, 1992
- Vol. IV: *Ancillaries, Plumbing and Sewerage*, 1992
- Vol. V: *Electrics in Buildings*, 1992.

Usually the operating aspects of the O & M manuals covers mainly plant and equipment usage procedures. So far in the history of BS development little has been contained in manuals regarding the instruction of building occupants in the use of systems at their workplace. It is an area that has been much overlooked. Few building occupants are aware of how their local environment is created, or the limits to which they can adjust it to personal choice. This is changing. The greater integration of building fabric and structure and services in the creation of low energy building brings with it an essential need to explain to the occupants how their building operates and what local scope they have in determining the environment they work in. The site manager should understand the content to be covered by O & M manuals. Drawn from BSRIA's *Handover Information for Building Services* TN 15/95 [5],

Table 11.2 The contents of an operation manual (Source: BSRIA TN 15/95, *Handover Information for Building Services*.)

A recommened strategy for operation and control
The procedure for seasonal start-up and shut-down
An outline of the general operating mode
Control data (location, effect, object, sequence, limits of capability, modes, set points)
Procedures and sequence for start-up, running and shut down, under both normal and emergency conditions
Interlocks between plant items
Operating procedures for standby plant
Precautions necessary to overcome known hazards
The means by which any potentially hazardous plant can be made safe
Target figure for both energy consumption and energy costs
Forms for recording plant running hours, energy consumption and energy costs
Procedures for the functional testing of safety systems

Table 11.3 The contents of a maintenance manual (Source: BSRIA TN 15/95, Handover Information for Building Services.)

GENERAL MAINTENANCE PROCEDURES

The manufacturers' recommendations and instructions for maintenance must be detailed for each item of plant and equipment installed. Clear distinction should be made between planned tasks (preventive maintenance) and work done on a corrective basis. Instruction should be given on each of the following, as appropriate:

- the isolation and return to service of plant and equipment
- adjustments, calibration and testing needed
- dismantling and reassembly
- the exchange of components and assemblies
- dealing with hazards which may arise during maintenance
- the nature of deterioration and the defects to be looked for
- special tools, test equipment and ancillary services
- procedures for the removal of heavy/bulky items e.g. plant mounted in the roof space

MAINTENANCE SCHEDULES

Maintenance schedules should be provided for all the preventive maintenance tasks identified under the previous heading of 'General maintenance procedures'. These should be based on both manufacturers' recommendations and other authoritative sources (e.g. statutory or mandatory requirements) and should include:

- examinations
- tests
- adjustments
- calibration
- lubrication
- periodic overhaul
- condition monitoring and assessment
- functional testing

Table 11.2 covers operation and Table 11.3 maintenance. Sections may also be added in either or both covering specific aspects of:

- fault finding
- summary schedule of lubrication for all plant
- modification information
- disposal instructions
- names and addresses of manufacturers
- drawing and chart schedule.

Associated with both O & M is the documentation usually referred to as Manufacturers' Technical Literature (MTL). Even quite small jobs attract an extensive bundle of such literature; it is now usual for these

Table 11.4 The MTL manual – information to be provided (Source: BSRIA TN 15/95, Handover Information for Building Services.)

Description of the product as purchased
Performance – behavioural characteristics of the equipment in use
Applications – suitability for use
Operation and maintenance details
Labour, plant, material and space resources required
Methods of operation and control
Cleaning and maintenance requirements
Protective measures
Labour safety and welfare associated with the equipment
Public safety considerations

to be retained in a separate manual cross-referenced to the O & M manuals. The Manufacturer's Technical Literature should provide the information listed in Table 11.4, also from the BSRIA *Handover Information for Building Services*.

11.2.2 FORMAT AND QUANTITY

A good test of how well O & M has been specified is to study the specification to see whether the designer has clearly spelled out how the information is to be presented. Any job over half a million pounds BS value will find itself very hard pushed to contain operating, maintenance and manufacturers' technical literature in one manual. Throw in the commissioning results and we soon find ourselves talking about books and volumes. Designers with a full appreciation of what it is they are asking for will clearly define the presentation along the following lines:

- Vol. I: Operating Manual
- Vol. II: Maintenance Manual
- Vol. III: Manufacturers' Technical Literature
- Vol. IV: Commissioning Records.

If the project is a multi-building site we could be in for four or five volumes for each building. Depending on technical complexity there may also be separate specialist manuals for:

- BMS (with or without HVAC controls)
- HVAC controls
- Security
- PABX
- Data
- Kitchens
- Cold rooms.

Of course, not all jobs are like that. For simple sites, e.g. social housing, student accommodation with central systems, one book with three sections will often suffice. Should the job be of domestic scale such as social housing then simple 'slim' manuals and perhaps a user leaflet (plastic encapsulated) will be adequate. Having given the site manager a warning as to what to look for and expect from the point of format we will now turn to quantity. If the manuals are all to come from one source the site manager will consider himself fortunate. The designers may have specified the appointment of a specialist technical author who will deal with the BS contractor(s) and produce everything to a specified 'house style'. The technical author may be employed in a number of ways. It may have been part of the employer's requirements for the PC to employ the technical author direct. More often than not if it is a job with a multi-service BS contractor then it could be part of their requirements to employ the technical author. Another route is for the designer to be the technical author and provide the documentation for a separate fee.

Multiple sources pose more of a problem for the site manager. Each BS contractor directly employed by the PC will have their own specified requirements for providing documentation. The size of the managerial problem is:

Number of sources × number of buildings × number of copies = quantity

11.2.3 PLANNING, PROGRAMMES AND APPROVALS

Look again at Fig. 11.1. It is all there on this backwards looking programme. From the approvals early in the job for samples of covers, volume depth, material, lettering, binding and contents dividers through to the cycles of approval for first and second draft on to handover. Two points to note. Allow four weeks for an approval cycle of issue and return. Certainly work to a tight fortnight (10 working days) cycle, but the logging in and posting out from and to originator and recipient will take the rest of the time even in the quickest teams. Be realistic; documents are very rarely approved first time round and you must allow for the worst case contingency of a third cycle. For the example shown in Fig. 11.1 of a one-year contract on an air conditioned building, three four-week cycles, say three months, is achievable, providing you have the front end approvals.

On large multiple building projects such as prisons and hospitals with many document sources all flowing through the PC on the approval route, a dedicated management handling resource is needed.

11.2.4 MNEMONICS

Mnemonics are a risk area for the site manager. These alphanumeric references for buildings services plant and equipment must be coordinated and project-specific rules laid down if problems are to be avoided. Inconsistencies manifest themselves in a number of ways. Consider the following situation with separate mechanical and electrical services contractors, and a separate BMS specialist. Take any motorized valve. The mechanical services firm may reference this in manuals and on record drawings as MV4 RP2 – motorized valve No. 4 in roof plant room 2; the electrical contractor having installed the field wiring may call it MV10LVL7 – motorized valve No. 10 at Level 7 (the level of the plant room); the BMS specialist will wish to continue using his own catalogue reference supplemented by a locational address, e.g. M3P8OF/A9L7, i.e. a magnetic activated, 3 port 80 mm flanged motorized valve on air handling unit 9, level 7. Quite rightly the frustrations of the end user are turned in bewilderment at a BS industry that provides such uncoordinated systems. The site manager can raise this subject early on in discussions by asking the simple question, 'are you all using the same plant and equipment referencing?' If the responses lead to a common system the site manager will have merited the thanks of a grateful client and saved some designers and installers from embarrassment.

11.3 Record drawings, wall charts and diagrams

11.3.1 UNDERSTANDING WHAT IS SPECIFIED

Record, as built or as installed – are they the same or different? Another fabled area of BS confusion. To unravel this confusion and to point the way forward with some clarity we will look at the usage of this terminology by referring to three documents published between June 1994 and March 1995. They are:

1. BSRIA TN 8/94, *The Allocation of Design Responsibilities for Engineering Services* [6];
2. the Association of Consulting Engineers *Conditions of Engagement 1995*, Agreement B (2) - for use where a consulting engineer is engaged directly by the client but not as a lead consultant [7];
3. the Construction Industry Advisory Committee, *A Guide to Managing Health & Safety in Construction* [4].

In the first of these references Appendix A. p. 29 defines a record drawing as:

> Drawing showing the building and services installations as detailed at the date of practical completion. The main features of the record drawings should be as follows:

- The drawings should provide a record of the locations of all the systems and components installed including pumps, fans, valves, strainers, terminals, electrical switchgear, distribution and components.
- The drawings should be to a scale not less than that of the *installation drawings*.
- The drawings should have marked on them positions of access points for operating and maintenance purposes.
- The drawings should not be dimensioned unless the inclusion of a dimension is considered necessary for location.

This correlates with the second reference taken from the ACE Document, clause I, Definitions, p. 14: 'Record Drawings – drawings normally prepared by a Sub-Contractor in order to provide the Client with a record of the Works as installed.' The final reference is taken from Appendix 4, 'The Health & Safety File', in CONIAC's *A Guide to Managing Health and Safety in Construction*, which lists in clause 2, 'Relevant information which could be included in the Health & Safety File'; 'Record or "As built" drawings and plans used and produced throughout the construction process.'

Later the same appendix states:

> 12 The provision of 'as built' and 'as installed' drawings is a common requirement in most contracts. Drawings are a good vehicle for the transmission of information between designer, contractor and back to the client. Drawings can also be a very good way of providing information required under the CDM regulations, particularly for inclusion in health and safety plans and the health and safety file.
> 13 The accuracy and usefulness of 'as built' and 'as installed' drawings varies in common experience. While absolute accuracy may not always be possible, attempting to achieve this will improve the provision of information. There can be difficulties in gathering all the information needed for accurate 'as built' and 'as installed' drawings. On large projects, it may be necessary to set up clear procedures to collect and validate this data. There may be many ways of presenting and storing this information. This is likely to develop over the coming years.

Some clarity seems to be emerging and the author dares to suggest that the terminology should be refined and applied:

- record drawings as defined by BSRIA in TN8/94 [6];
- as built and as installed where greater accuracy of information is required in part or whole, the extent of which is a specified requirement.

Fortunately, no such difficulty exists with wall charts and diagrams providing the designer clearly states what is required, i.e. wood framed, under glass or encapsulated in plastic, etc.

Until there is a wider use of clearer definitions of drawings to be provided at handover, the site manager is advised to find out what is called for on his specific project. This can be done by forming his own opinion of specified requirements, and following that up by asking the BS contractor at the pre-award meeting, 'What have you allowed for?' Certainly it will ease any end of job difficulties if requirements for drawings can be clarified at the beginning of the project, and be properly costed, thereby avoiding scope for a claim.

If the contract has determined that dimensionally accurate information is to be provided on the handover drawings, and this may be necessary on jobs such as prisons, hospitals and MOD establishments, the project programme can be affected. Time must be allowed on the BS contractors' programmes for the surveying of constructed services before their enclosure. Beware of the BS contractor under pressure on a site with a number of identical buildings who only surveys one. The buildings will not be identical. Variations in construction tolerance of the building elements and minor differences in tradesmen's practices, particularly, for example the conduit routes, makes each building different. A good specification will state the requirements for drawn information along the lines of Table 11.5.

11.3.2 PLANNING, PROGRAMMES AND APPROVALS

This follows the strategy of Fig. 11.1 and the comments of section 11.2.3.

Table 11.5 The specified requirements for record drawings, wall charts and diagrams, etc.

The specification should state the requirements for:
Level of information, i.e. all architectural and services construction details omitted
Scales to be used
Title block, layout and logos: samples required
Full sized drawings and reduced
Negatives and/or prints, microforms
Standards of finish, e.g. framed under glass, encapsulated in plastic, for plant rooms, and panel hung drawings, schematics, diagrams, distribution board schedules: samples required
Number of copies, negative plus 'x' prints
Data transfer, floppy disk

11.4 Familiarization, instruction and training

11.4.1 GENERAL

We enter another potentially risky area of activity. Risk will manifest itself through the vagueness of the specification of the designer or under-estimation by the BS contractor in complying with a good specification.

Looking at extremes, 'vagueness' is the clause that states, 'Instruction shall be given in the operation of the mechanical services.' It may be repeated changing mechanical to electrical in that specification, and omitted completely from the public health section. It has even been known for large specialist installations, for example of kitchen and laundry equipment, to have had instruction and training omitted from their specification, presumably, assuming that the operatives will be 'expert'. This is not so. On those two specialist installations it is not uncommon for the majority of operatives to be untrained or at best semiskilled.

The better specification will describe what is to be provided, to whom and for what duration. It will also describe the expected levels of expertise for both the trainer and the trained. The best specifications will probably describe a three level process in ascending order of intensity:

1. familiarization;
2. instruction;
3. training for;
   - general building services – operation and basic maintenance;
   - specialist services – operation and basic maintenance;
   - process services – production (hot/cold start) and maintenance.

Each of the three main levels have different but related, sometimes overlapping, aspects of how, when, where, by whom and for what duration.

11.4.2 FAMILIARIZATION

This can be a geographical walk around the engineering services systems, particularly describing the purpose of plant and equipment, given perhaps over one or two days, according to the density of services, numbers of buildings and their size. The walk round should inform the management e.g. facilities and premises managers, supervisors and operatives with O & M responsibility. Familiarization should point out what plant and equipment is manual or automatic in its operation, and refer recipients to the O & M manuals and record drawings. It may be specified as starting with a 'teach in', pointing to the available documents and describing what will be found in them.

Familiarization must be planned and given by persons with good job knowledge who are capable of talking in front of people to impart confidence in the services they will be responsible for.

11.4.3 INSTRUCTION

The specification should describe the level of information to be imparted to staff operatives about the operation of the engineering systems in normal, emergency and standby mode. Some instruction in basic maintenance such as filter cleaning and changing, belt tensioning, location and entry into plant, e.g. air handling units via access panels may have been specified. The extent of demonstration for both operation and maintenance should be specified, together with the anticipated duration of the instruction and the level of competence the recipients will be expected to have. From the above information the provider of the service can assess their own and specialist resources required.

In his management role the site manager should call for a meeting between the providers and recipients of instructions to review the abilities of both before instruction commences. Unfortunately, left to their own devices, both parties may start accusing the other of inadequacies. A view from the higher level authority of the site manager's position can do much to bring the parties together at the right level of competency. The site manager may form a judgement that the recipients need more training, possibly outside the scope of the contract, before they will benefit from the instruction to be given by the BS contractor. The site manager may also call for improvements to be made in the level of competency to be provided by the BS contractor.

11.4.4 TRAINING

Training is a specified level of trade skill to be given to staff and operatives assigned by the end user. The training should have emphasis on achieving the skill levels needed to carry out basic maintenance. Training could also cover enhancing basic skills by upgrading operatives to maintain specialist plant and equipment, e.g.:

- filters trained up to maintain boiler and refrigerant plant;
- electricians trained up to maintain HV switchgear, generators and UPS systems;
- plumbers trained up to maintain water treatment standards;
- supervisor trained up on the BMS system.

Specialist training will require the identification of suitable staff and operatives by the end user, and appropriate courses by the designer, BS contractor or specialist subtrader. In addition to a specialist BMS supplier training course, the designer may specify some 'on-site' post-contract training to be given.

Process training will almost always be the responsibility of the end user but some processes require careful interfacing, e.g. continuous hot process dough ovens must be demonstrated 'hot' and handed over to user, avoiding cooling down which could damage refractories.

11.4.5 WHAT WILL PROBABLY BE SPECIFIED

The most likely description will be the specification of something from each of the three levels, perhaps in the following way:

1. familiarization for:
   - management;
   - facilities manager;
   - maintenance manager;
   - supervisors; and
   - leading operatives
2. instructions for supervisors and operatives, preceded by
3. training for BMS supervisor/manager and lead operatives.

All of the above may not be clearly stated in the specification and the site manager can do much to bring the parties together to refine responsibilities to ensure that what is required is delivered on time for handover.

11.5 Spares, keys, tools, media and replacements

11.5.1 GENERAL

With O & M manuals and record drawings, instruction and training well under control, risk has been tamed. Or has it? The provision of spares, keys, tools and media can provide one last painful sting before handover. By now the site manager is sick of being told 'Understand what is specified.' What we mean is assure yourself that your BS contractor is aware of the needs of this area. Less ambiguity surrounds these provisions as quite clearly they relate to the 'in place' works. Of course, there is a caveat – there always is. The specification that states, 'Provide spares as recommended by the manufacturer', takes us into the area of trying to define the difference between spares which are known to be required to replace parts that wear to particular time frames, and parts the manufacturer would like you to have because he sells them. Back to basics, it is the responsibility of the designer to agree with the client a policy for the provision of spares and clearly define the requirements in the specification. The site manager therefore must ask his BS contractor 'Are you happy that the specification of spares, keys, tools and media is clear?' Into this category of loose collectibles to be handed over comes any instrumentation requirements from the commissioning section of the specification. Portable temperature and humidity readers, noise meters and data loggers, etc., etc. are not unknown.

Meeting the requirements does not require the same depth of planning and approving, but on a small job taking a close look at what is required should start four weeks before handover. On a large job you may find that spare motors are required to be held in stores and their delivery period may be four to six weeks from order. The earlier the subject is looked at the more comfortable the site manager will feel about his BS contractor's capability.

11.5.2 SPARES

The level of provision may well depend on the end user's policy for maintenance, i.e. in house or contracted out. It will also depend upon the function of the buildings. Buildings functioning at the forefront of health care, security and high volume finance transactions demand systems of high reliability. Mainly this is covered by the designer providing duplicate plant and equipment with appropriate levels of redundancy. Whatever level of standby and emergency capability is provided, any outage for these buildings is a serious matter and spares must be speedily available from on-site stores or via the term maintenance contractor.

Spares frequently specified to be held on site are motors, fuses, drive belts and gaskets.

11.5.3 KEYS

A subject in which site managers are well versed; so suffice to say they need only examine the BS contractor to make sure he has keys for control panels, switchgear, fire alarm panels, lifts, valve padlocks, valves, e.g. lockshield not forgetting the humble radiator air vent keys.

11.5.4 TOOLS

In addition to what the designer specifies for tools, plant and equipment can arrive on site and be erected and commissioned by specialist contractors. It is not unknown when after handover and a key piece of equipment has failed that the maintenance staff discover from the manual that a specialist 'deep reaching' tool should have been provided to facilitate a simple mechanism release that gets the equipment operational again. Telephone/fax to the manufacturer (and they won't always speak or understand English) determines after frustrating delay that the tool was left by the fitter inside the right hand end cover! The message for the site manager is once again ask your BS contractor to trawl the manufacturer's technical literature and list any 'special' tools required for the basic operation and maintenance of plant and equipment.

From the example above it can be seen that this is a pitfall into which it is easy to fall unwittingly. Jobs can be signed off in good faith, but ignorance of 'special' tool requirements go undiscovered.

11.5.5 MEDIA AND REPLACEMENTS

Bracketed with greases, oils, water treatment chemicals, ventilation and air conditioning system filter media, we will couple the replacement of lamps, tubes and other items of the permanent systems run under contractual agreement prior to handover

Very crudely the fall off in levels of illumination, decay in potency of water treatment and the dirt loading of filter media are related to the time between turn on, addition, and commissioning respectively. The client expects the installation to be handed over in pristine condition. Through the specification the designer seeks to achieve that objective. If the specification prohibits the use of the permanent installation other than for testing and commissioning, after which they are to be turned off, the site manager may seek a waiver to the contract conditions. Every site manager is aware of the difficulties of trailing temporary electrics for power and lighting through near complete building work. Door edges and frames, floors and skirtings get scuffed adding to the difficulties of builders and final clean. It is difficult to clean a building without permanent lighting. This is but one aspect of site productivity. All site managers will have experienced the increase in productivity of building services and finishing trades when permanent lighting is on. There is a need for both client/DT and the PC to come together to agree on the use of permanent systems and negotiate sensible cost adjustments for their use. After all there comes a point before the end of the contract when permanent lighting, HVAC and all general services systems must go live. What client and DT want to conduct their final inspections through a temporary lit building waymarked by a hazardous trail of cables – silly isn't it? The other side of the situation is that no client/DT can condone a PC who is profligate in the use of energy and water, etc., for which the client has signed up supply agreements with the gas, water and electricity companies.

Equitable negotiation can only come from both parties understanding the other's needs. At present there is an imbalance of understanding to be addressed by most PCs.

11.6 Maintenance

11.6.1 COMPREHENSION

Some PCs still believe that the defects liability period is that time during which the BS contractor must maintain the systems. Not so. Possibly the BS contractor was asked to submit a price for maintaining the installation for 12 months after handover, but most usually this becomes an agreement that is quite separate from the main subcontract.

Generally BS contractors have the same contractual obligations as the PC. Although on small jobs it is not unknown for the builder to have a six month defects liability period and the services firm one of twelve months. We know this is done so that the HVAC systems will have operated through all of the climatic seasons and any installation defects arising from the changed operational patterns can be attended to. The difference in defects liability period between PC and BS contractor gives the former difficulties in rendering the final account.

Depending on how the specification has set up maintenance, the post-handover period may involve the site manager. If a planned preventative maintenance system has been specified to be provided by the BS contractor then the PC is involved in making sure that handover requirement is achieved. Alternatively, if the BS contractors' tender for maintenance has been accepted, the PC should only be involved to ensure that his contractor does not cut the corners of familiarization, instruction and training because he feels he will be maintaining the plant and there is little point in instructing his own resources. Any reduction in the training and instruction programme may involve a clawback being negotiated by the QS.

11.6.2 PLANNED PREVENTATIVE MAINTENANCE

It can be reasoned that planned preventative maintenance is covered in the periodic schedules of servicing contained in the maintenance manuals – see section 11.2. In this section we refer to it in its two computerized guises. In both these forms it can be used in addition to, or instead of the 'normal' O & M format.

Most BMS firms can offer computerized planned preventative maintenance varying from the simple captured text of the maintenance manual through a variety of forms up to the latest dynamic systems of Table 11.6. Stand alone systems were the first generation of computerized PPM systems. The systems were prepared by experienced maintenance managers from the National Health Service and petrochemical industry. These produce electronically generated hard copy in the form Table 11.7.

The integrated system is more suited for those projects where the maintenance will be carried out in house supported by a few specialist

Table 11.6 Coverage of BMS integrated PPM

Asset registers
Periodic maintenance scheduling
Labour resource optimization
Fault diagnosis and attendance grading
Maintenance log printout – resource times and material needs
Maintenance and fault records
Maintenance and fault analysis and trend logging
Costings

Table 11.7 Coverage of a stand alone PPM

These were the first generation of computerized PPM systems. The systems were prepared by experienced maintenance managers from the NHS and petrochemical industry producing:
Asset registers
Periodic maintenance schedules
Labour resource optimization
Job cards
Record cards

term contractors. The stand alone system is the type of arrangement created by term maintenance contractors from the hard copy of the O & M manuals. It is of less interest to the PC, whereas the BMS integrated system will need to be set up and running for handover day plus one. The site manager will come across the term 'condition monitoring' and may find this is incorporated into the BMS integrated programme of maintenance. The practices of condition monitoring are just one in the study and reaction to wear and tear which is known as terotechnology. Condition monitoring is moving from a form of research into a reliable maintenance management tool whose application in the BS industry is on the increase. Applied to moving machinery in the monitoring of, e.g. shaft vibration and bearing wear, its objective is to forestall unacceptable breakdowns, prolong life and make the timing of major cost expenditure more predictable.

Condition monitoring is expensive and will need careful application to provide the best value. We can see its use on generators, rotary UPS systems, refrigeration and air handling units plant serving critical operational functions such as trading floors, computer rooms, operating theatres and continuous process industrial plants.

11.7 Post contract

11.7.1 AFTER HANDOVER

Not for long does the site manager float free after handover before the fax and telephone notify some fault of varying magnitude. Section 9.8 dealt with the post contract commissioning activities of system proving and fine tuning. Here we will look at fault analysis and rectification.

11.7.2 FAULT ANALYSIS AND RECTIFICATION

This will be a simple starting point approach from which, according to the nature of the problem, the site manager may need to call upon assistance from a wider field. This may be found in Chapter 12.

The resolution of a great number of faults will present no problem to the PC. They will be latent problems in materials and workmanship undetected by the inspection processes, manifesting themselves through the stresses and strains of dynamic systems called upon to function for longer periods than during commissioning activities. The difficult ones are those that do not respond to initial remedial action where the fault was not so obvious as first apparent.

In the investigation and rectification of any fault the site manager should be prepared for it not to be as simple as it seems. It is essential in those circumstances to have a strategy to put in place and Table 11.8 is suggested as an operation framework for fault analysis. This will be expanded upon in Chapter 12. Remember in any fault investigation the search for causes should be divergent to the point of lateral thinking. After the cause has been found the ways and means of its resolution should be convergent.

It is imperative that accurate records are kept of defects attended to

Table 11.8 Fault analysis procedure (Source: BSRIA.)

Failure: the termination of the ability of an item to perform a required function (BS 4778)
Don't panic
Limit the damage
Establish the history of the system – design, installation, commissioning, O & M
Compare with design – variation, modification and additions
Check working order – controls
System contents – contents and environment
Conclusions – identification, remedial action
Call in the experts

Table 11.9 Checklist for building documentation (Source: CIBSE TM 17.)

Fire certificate, showing fire compartmentation, etc.
Records of fire detection and alarm tests showing test point used for each test, dates and smoke detector tests
Sprinkler systems test records
Smoke extract system test records
Escape route pressurization system test records
Emergency lighting system test records
Fire extinguisher and fire hose reel test records (in accordance with BS 5306)
Legionella risk assessment (in accordance with ACOP 1991, Regulation 6 of Control of Substances Hazardous to Health Regulations 1988 (COSHH), the Health and Safety at Work, etc. Act 1974 and HS(G) in the Health and Safety General Series)
A register of compliance with the Notification of Cooling Towers and Evaporative Condenser Regulations (1992)
Records of *legionella* risk management implemenation for at least past two years (Approved Code of Practice 1991, COSHH 1988, Health and Safety at Work, etc. Act 1974)
COSHH records
Lift insurance inspection reports
Lifting equipment insurance test reports and certificates
Pressure vessel and system test reports and certificates
Fume cupboard test reports and certificates
Operation and maintenance manuals with sections dealing with how to isolate equipment and emergency procedures (Health and Safety at Work, etc. Act 1974)
Electrical earthing and insulation test records (every five years, in accordance with BS 7671)
Portable appliance test records (Electricity at Work Act)
Waste disposal and handling procedures (Environmental Protection Act)
Noise assessments
Asbestos awareness report
General risk assessment
Asset register and installation record drawings

during the defects liability period. Obviously it is in the interests of the PC and BS contractor that they agree with the client/end user what details are recorded, e.g.:

- the nature of the defect;
- any consequential damage to other services or building elements arising from the defect;
- remedial action taken and by whom;
- date and time of event and related activities.

Unlike many elements of building works where defects can be held over to a later date in the defects liability period and dealt with in an economical manner, the BS contractor is not so lucky. Once again the dynamic nature of building services dictates, more often than not, that response to a problem must be immediate in order to secure the building function and maintain safe occupancy for all.

11.8 Information for the building owner and user

Not all documentation that must find its way into the PC's health and safety file, for onward transmission to the PS and ultimate handover to the end user, will be contained in the O & M manuals and record drawings. We will therefore end this chapter with Table 11.9, from the Technical Memorandum Appendix B, *Building Services Maintenance Management* [8]. The table 'is a list of documentation that, where applicable to the building concerned, building owners must keep. Other documentation may be desirable but does not arise out of any statute or regulation. All the documents are to be kept in the building and must be freely available for inspection.' This prefacing comment to the table, published by CIBSE prior to the CDM regulations coming into force, is recommended to the site manager for inclusion in the health and safety file. By extending Table 11.9 by two columns the site manager can create a project specific responsibilities matrix for completion in conjunction with the BS contractors. The first column can identify the provider of the requisite documentation and the second column its location, i.e. contained with the O & M and commissioning manuals or presented loose.

References

[1] *Health & Safety at Work Etc Act 1974*, HMSO, London.
[2] *CDM Regulations 1994* (1995) HMSO, London.
[3] Construction Industry Advisory Committee (CONIAC) (1995) *A Guide to Managing Health and Safety in Construction*, HMSO, London.
[4] *HVCA Standard Maintenance Specifications* (1990, 1992) HVCA, London.
[5] BSRIA (1995) *Handover Information for Building Services* TN 15/95, Bracknell.
[6] BSRIA, *The Allocation of Design Responsibilities for Building Engineering Services*, Technical Note TN8/94, Bracknell.
[7] The Association of Consulting Engineers (1995) *Conditions of Engagement 1995*, Agreement B(2), London.
[8] CIBSE (1994) *Building Services Maintenance Management*, TM 17, London.

Help yourself 12

12.1 Information plus knowledge equals compliance

12.1.1 INFORMATION AND KNOWLEDGE

Information is not knowledge. Its value is limited by individual, team or corporate knowledge. The stocks and shares listings of the *Financial Times* mean little to the reader without knowledge. Information is therefore only knowledge if it is capable of interpretation by the recipient and adds value to his other knowledge. An enquiry for a construction project is 'given' information. The issuing authority has imparted their knowledge in the form of information. Some tenderers return enquiries explaining, 'We are not in this area of business.' Others, seeing the enquiry as being on the fringe of their field of work, while recognizing the extended risk will still submit a bid. Most tenderers confident of their knowledge will interpret the enquiry, and its implied or explicit requirements to comply with legislation, standards, codes and guides, into an intended compliant bid. They may apply their knowledge further and put forward alternative offers for the client's consideration.

The contract will proceed with more information, e.g. contract drawings and variations, issued under cover of an instruction from the client/DT via the contract administrator. As work proceeds clarification is required to issued information. It may be ambiguous, incorrect or have different meanings according to the reader's knowledge level. It is the continuous ebb and flow turning information into knowledge, back to information, and on to enhanced knowledge that ensures specified requirements are met.

12.2 Compliance with laws, standards, codes and guides, etc.

12.2.1 GENERAL

All work must comply with the legislation of the country within which it is being constructed. For the UK that means complying with statute law laid down by Acts of Parliament. This law is enabled through regulations. Some establishments such as the Crown Estates and certain areas of MoD activities do not have to comply with all legislation and are granted immunity from prosecution. No doubt the same situation exists in other countries.

It is not necessary for all of the legislation with which a construction project must comply to be referred to in the contract documentation. After all, to comply with the law that bounds your business you must understand that law. This is simply stated, but it is easy to move quickly from black and white into the grey. Take the building regulations in the matter of fire stopping. A designer must know the requirements and should specify one of the 'deemed to comply' methods for sealing holes around building services penetrations through structure and fabric. If the designer is specific then the builder or BS contractor, if it is in his works, only has to meet that specified requirement. Should the designer have made the incorrect selection it is obviously his fault. If, however, the designer has left the choice of fire stopping to the builder or his BS contractor then in that delegated area of 'design' responsibility they need to know all of the compliant alternatives that will meet the regulations.

According to the function of the building the designers will need to specify particular legislation with which the project must comply. If you are building a meat or poultry processing factory then the designers will have to comply with the requirements of the Food Safety Act 1990 and particular requirements of The Meat Products (Hygiene) Regulations 1994. These will be stated in the specification and the builder and BS contractor should certainly acquaint themselves with the requirements of the legislation, for they affect the way in which the building is constructed, commissioned, cleaned and handed over. The constructors may even need specialist help at tender stage in interpreting requirements of the legislation so that they do not default during construction.

Much legislation is empowered by Parliament having received a Directive, e.g. the Construction Products Directive, from Brussels. There seems an endless expansion of legislation, the criticism of which is deflected by politicians pleading not the Fifth Amendment but the EEC. Returning from our travels in Community countries we bring back stories of their apparent non-compliance in food hygiene, fire regulation and construction safety. But who would want to see the annulment or watering down of, for example legislation concerning Personal Protection Equipment (PPE), COSHH and CDM regulations.

To conduct their business the PC and BS contractor must be up to speed with the currency of their knowledge. Legislation is but one part. They must understand the way in which standards, codes, guides and related technical memoranda and technical notes are created and how to get hold of them when they apply specifically to their projects.

12.2.2 THE NATURE OF REFERENCE INFORMATION

The reader will pardon the fuzzy logic of the oversimplification which follows. The reader should be able to create his/her own sharp edges to the interpretation and keep him/herself afloat on the grey seas of project situations:

Legislation

All statute law is made by Act of Parliament. Common law is established by decisions or judgments in courts of law and has evolved over the years.

Regulations

These are the enabling instruments of the Acts. This legislation *must* be complied with. Failure to do so is punishable under the law.

Standards

A hierarchy of organizations set standards. The main group we will bracket as:

- International Organisation for Standardisation (ISO)
- European Committee for Normalisation (CEN)
- European Committee for Electrotechnical Standardisation (CENELEC)
- British Standards Institution (BSI).

These are the international, European and UK standards. In a few cases the applicable standards are the same, as in the triple numbered Quality Management System Specification, ISO 9000/EN2900/BS5750.

As common standards flow through from Europe BS standard numbers are changed and given EN numbers.

Remember – these standards are generally considered as minimum. The standards and practices of other organizations may be even higher and considered more of a benchmark.

Other standards

Research organizations, professional institutions, trade associations, universities, government departments and corporate bodies produce their own standards. These too are recommendations and will be embodied into contract documents on the route from client to designer to PC and specialist contractor, and onwards to suppliers and manufacturers. In reverse order manufacturers, suppliers, specialists and general trade contractors will, or should, apply their standard industry norms to meet but not take precedence over the specified standards.

Generally standards tell you what to do. Some may find that

statement debatable, but that is only because of their knowledge level. In reading the standards they draw upon their skill and experience. If they have not acquired the level of experience which allows them to understand the standard they will need guidance to implement it.

Codes

Codes mostly originate from the same sources as standards. Those coming from the higher levels of standard setting authorities are given the title, Approved Code of Practice (ACOP), e.g. L54, *Managing Construction for Health and Safety* related to the CDM regulations and was issued by the Health and Safety Commission.

Codes vary in their nature from documents which clarify requirements and in doing so edge from 'what to do' towards 'how to do it'. Technical codes generally are skeleton documents, i.e. they have a structural framework to them.

Guides

Guides are usually drawn up by the source organizations for standards and codes. Guides flesh out the codes providing them with the 'how to do it', e.g. *The Guide to Managing Health and Safety in Construction* prepared in consultation with the Health and Safety Executive, by the Construction Industry Advisory Committee (CONIAC). Another example is BSRIA's Application Guide AG2/89.1, *The Commissioning of Water Systems in Buildings*.

Technical memoranda and technical notes

These are taken together, usually from the same sources as guides, codes and standards. They may also go under other names such as bulletins, digests and newsletters. For BS we will take the BSRIA definitions, while recognizing they will not apply to all sources.

- **Technical notes** are reports prepared as a result of sponsored research and other technical work.
- **Technical memoranda** are occasional publications on diverse topics prepared from the deliberations of various specialists.

Bibliographies

A number of originating sources of standards, codes and guides, TMs, TNs, etc. also produce bibliographies in the form of publication catalogues general to their sphere of activity. You may also find that some organizations, particularly BSRIA, produce subject specific bibliographies.

Technical papers

These are available from the conference papers and technical journals of the generic sources already mentioned. These provide information that ranges from 'what to do' through to 'how to do it' and 'how we did it' on this or that job.

Directories

The industry is blessed with an excellent range of directories covering such areas as standards, organization membership, certification and registration of competency by companies, individuals and products, e.g. registered firm to BS5750, the Heating, Ventilating, Airconditioning and Refrigeration Register for operatives, and fire detection devices complying with LPCB Certificate 126a respectively.

12.2.3 SOURCES OF INFORMATION

From an understanding of the nature of information related to building services the PC and his site manager should be able to provide direction to others, and lead where necessary, in the resolution of problems. Involved as he will be the site manager should be able to drive others towards the acquisition of the best subject information, question or bring the parties together on its interpretation in the context of his project. Questions such as:

- 'What is the specified standard?'
- 'Is a current copy of that standard on site?'
- 'In the absence of a specified standard what authoritative guidance is available to us?'
- 'Is that guidance on site?'
- 'What parts of the standard or guidance are we having difficulty with?'
- 'Is there a higher level of knowledge on this subject available in your organization?'
- 'Have you researched whether this situation has occurred previously in the experience of yourselves or the industry?'
- 'How was that problem resolved?'
- 'Is it relevant to this job?'
- 'Were any external experts involved', if so

- 'Who were the experts?'
- Etc., etc. . . .

Tracking down what is known about the subject 'difficulty' should assist in defining the limits of current knowledge. In this BSRIA's *Information Sources in Building Services* [1] will be found most useful. This provides contact details for organizations and a subject listing. Appendix M has been developed from this to provide a listing of the range of information services available from professional institutions, consultancy associations and learned societies.

12.3 Dealing with faults and failures

12.3.1 FAULTS NOT DEFECTS

All this checking of information, clarified and refined by knowledge is fine for the static construction condition of building structure, its fabric and services systems during construction. Work is inspected, defects raised and cleared, the external inspectorate comes and goes. It is in the dynamic mode during commissioning, system proving and defects liability period that BS failures manifest themselves.

What we are looking at here are not defects as defined in section 10.1.2, or total failure. We are concerned with a partial failure or fault in performance, the resolution of which we will take through an analysis procedure and the occasional need to call in an expert.

12.3.2 FAULT ANALYSIS PROCEDURE

This is an exercise in the iterative process of information converted to knowledge and back into information, in problem resolution. The procedure is based on ideas set down by BSRIA quite a few years ago and which the author has found a helpful approach to trouble shooting.

- **Don't panic** – unless safety is involved do not destroy the evidence. Wrong conclusions may be drawn. If the subject failure presents visual evidence and time allows, then photograph it.
- **Stop further damage** – despite the warning not to destroy evidence there may be certain actions which must be taken promptly to limit the consequential damage.
- **History of the system** – collect all known details from design, installation, commissioning, operating and maintenance. A chronological history of the problem is invaluable. Note any decisions of political expediency that appear to impact the problem. However, be wary of information born of vested interest.
- **Compare with the design** – question variations from the original design, check on modifications and additions.

- **Working order** – make sure the system is working in accordance with design. Automatic controls may be stuck and system performance could be totally different to that expected. Controls in one part of the system may struggle to operate and compensate for failures to controls in other parts.
- **System contents** – check that the application of plant, equipment and components and operational environment are not different from that for which the system was designed.
- **Drawing conclusions** – can the cause of the problem be identified from the information collected? Is it a typical failure which could be expected?
- **Call in the experts** – you may want the problem investigated by experts or decide that certain features need scientific services. This is what experts are for.

  Remember the majority of faults will arise from one of the following sources:
  - incorrect selection of materials or finishes
  - design errors
  - installation inadequacies
  - commissioning, operating or maintenance malpractice
  - incorrect or inadequate client brief.

12.3.3 ASSIGNING RESPONSIBILITY

In order to assign responsibility for a failure one must first understand the responsibilities of the parties to the project. It is not unusual for both designer and the installer to prepare reports of their investigations which may lead the site manager to discover that they are incompatible. If they both blame a supplier or manufacturer the situation is easier for it is down to the BS contractor to seek remedial action. If stalemate persists between designer and installer it may fall to the site manager, in order to break a deadlock, to seek agreement to the appointment of an external expert.

12.4 Calling in experts

The resolution of major problems, particularly those concerning system performance, are difficult to deal with. Occurring towards the end or after the job has been handed over no one in the chain has any money to put it right. Polarization is inevitable as the site manager witnesses a flow of correspondence and 'expert' reports angled from each party's vested interest. The PC and site manager are unavoidably involved. The contract is invoked and instructions flow from the client/DT which may not be acted upon despite '7-day notices'. In picking the moment to strengthen the intervention, and control draining expenditure, is not easy for the PC. In any such dispute the site manager must always consider that a moment in time may arrive when he has to spend his company's money in a controlled way, in order to solve a problem belonging to others, which is burning off his profit with seemingly no chance of its recovery. For the last time it is necessary to refer to the differences in contractual arrangements and the nature of building services when compared to the static elements of the construction works.

PMs, D & B contractors and, to a lesser extent, construction management contractors can and must be responsible for preventing debate on differences sliding into acrimonious dispute. The management and traditional lump sum contractors may find their influence is atrophied by being 'the piggy in the middle'. Nevertheless all may find they are recommending that an expert is called in.

If, for whatever reason, you as a project manager or PC are assigned or contractually bound to take the lead in appointing an expert, then get the best. Regrettable though it may be, the PC should always be mindful that the expert may have to appear as an 'expert witness' in arbitration or some higher level legal proceedings. It may not be so much as a dispute over the expert's technical findings, but the strength of views held on the contract conditions that will cause the legal profession to wet the fiscal appetites. The special appointee must therefore be an expert in the subject field, comfortable in writing reports and appearing in court. A presence and equable temperament is important. It should not be too difficult to determine the nature of the problem to be investigated although the complexity of BS engineering in its mix of mechanical, electrical, electronic and chemical elements can on occasion give rise to the need to appoint more than one expert to deal with the constituent aspects.

As most disputes seem to occur between designer and installer or installer and manufacturer/supplier there is no merit in appointing another 'general' BS design consultant. If you do you may feel that their findings are suspect. Naturally they will not wish to find against a fellow consultant, which leads to the installer, manufacturer and supplier saying unhappily 'Well, what do you expect from another Designer?' Unfair though this may be it has not helped to resolve the situation. The reverse situation also applies with some undoubted experts on 'retainers' from their past employers in manufacturing. Their natural bias is towards the product. Nevertheless some truly independent experts do exist in the specialist fields of:

- HVAC
- comfort environments and building science
- refrigeration
- vertical transportation
- laundries
- kitchen installations

- security
- combustion
- electrical motive power
- information technology.

In addition to the independent experts who are sometimes difficult to locate there is the specialist knowledge of the industry research organizations, e.g. BRE, BSRIA, CIRIA, etc. and of course the universities.

When you have tracked down one or two potentially suitable experts, check them out. We know time is short and the problem must be resolved yesterday. But, the expert you are going to appoint will need to have a success rate in the high 90s, whose findings will not be 'rubbished', and for this reason it is worthwhile spending some time on a pre-qualification process. Consider:

- Do they understand the problem and have solved similar issues previously?
- Are they just keen to get involved because the problem interests them academically?

Check out what information they are going to need. Can a start be made while the information is being compiled? Ask them how long they will need to investigate and prepare their report. Experts are not cheap but reflect the value to the project. Experts must be carefully briefed, but don't tie them down. You are not the expert and should not tell them in fine detail what it is that they should be looking at. Restrict the briefing to technical matters. If they are asked to look at the contractual arrangements and form views on the political atmosphere of the project you will be asking them to go outside the realms of their expertise. It will also prolong the presentation of their report. Contractual and political views could be non-expert opinions, 'torn to shreds' in the courts as a way of belittling their views on technical matters.

When experts start out on an investigation they may for all their expert knowledge come up with conclusions that will recommend further avenues of investigation. Before proceeding, discussions should take place and views expressed as to whether or not an ultimate conclusion is likely to be achieved and at what cost. Hard though it may be to understand, phenonoma do occur in our buildings, not only on BS but with structure and fabric, the causes of which are only understood following investigations. Fortunately building services phenonoma are usually not dangerous, if we accept that they only partially contribute to sick building syndrome. Take the incidence of harmonics in a project's electrical infrastructure as one such phenomenon. Existing as they do in all jobs it is only when harmonics occur to an unpredictable level that problems arise.

End note

Buildings with the aid of BS defend the occupants and processes from the vagaries of the external environment and enable the functions for which they were designed to be performed safely. When the continuous presence of quality is absent in part, or whole, it is usually manifest in some level of BS failure. The boiler didn't come on, the chiller packed up, the fuse has blown, the sprinkler failed, the lift is broken, etc., etc. Frustrating but mercifully not often life threatening, their impact is nevertheless more noticeable than cracked floor tiles, peeling paint and sticking doors. Comparing like with like is not possible, but BS probably do not fail any more than buildings. When the latter fail the consequences can be horrendous, as a read of *Why Buildings Fall Down* [2] graphically explains. Generalizing, when BS fail it can be disruptive to a wide degree, when buildings (structures) fail it's frightening. Both risks can be reduced by using expertise in design and construction and good management of the whole process, not in investigating what went wrong. For all, the advice of William LeMessurier, doyen of Boston Structural Engineers as quoted in [2] is given, 'Any time you depart from established practice make ten times the effort, ten times the investigations. Especially on a very large scale job.'

References

[1] BSRIA (1995) *Information Sources for Building Services Professionals*, Bracknell.
[2] Levy, Matthys and Salvadori, Mario (1992) *Why Buildings Fall Down*, W.W. Norton, New York.

Appendix A CAWS – Common Arrangement Work Sections. R–X including first and second level descriptions

R	**Disposal systems**		
R1	Drainage	R10	Rainwater pipework/gutters
		R11	Foul drainage above ground
		R12	Drainage below ground
		R13	Land drainage
		R14	Laboratory/Industrial waste drainage
R2	Sewerage	R20	Sewage pumping
		R21	Sewage treatment/sterilization
R3	Refuse disposal	R30	Centralized vacuum cleaning
		R31	Refuse chutes
		R32	Compactors/Macerators
		R33	Incineration plant
S	**Piped supply systems**		
S1	Water supply	S10	Cold water
		S11	Hot water
		S12	Hot and cold water (small scale)
		S13	Pressurized water
		S14	Irrigation
		S15	Fountains/Water features
S2	Treated water supply	S20	Treated/Deionized/Distilled water
		S21	Swimming pool water treatment
S3	Gas supply	S30	Compressed air
		S31	Instrument air
		S32	Natural gas
		S33	Liquid petroleum gas
		S34	Medical/Laboratory gas
S4	Petrol/Oil storage	S40	Petrol/Oil – lubrication
		S41	Fuel oil storage/distribution
S5	Other supply systems	S50	Vacuum
		S51	Steam

S6	Fire fighting – water	S60	Fire hose reels
		S61	Dry risers
		S62	Wet risers
		S63	Sprinklers
		S64	Deluge
		S65	Fire hydrants
S7	Fire fighting – gas/foam	S70	Gas fire fighting
		S71	Foam fire fighting

T Mechanical heating/Cooling/Refrigeration systems

T1	Heat source	T10	Gas/Oil fired boilers
		T11	Coal fired boilers
		T12	Electrode/Direct electric boilers
		T13	Packaged steam generators
		T14	Heat pumps
		T15	Solar collectors
		T16	Alternative fuel boilers
T2	Primary heat distribution	T20	Primary heat distribution
T3	Heat distribution/utilization – water	T30	Medium temperature hot water heating
		T31	Low temperature hot water heating
		T32	Low temperature hot water heating (small scale)
		T33	Steam heating
T4	Heat distribution/utilization – air	T40	Warm air heating
		T41	Warm air heating (small scale)
		T42	Local heating units
T5	Heat recovery	T50	Heat recovery
T6	Central refrigeration/ Distribution	T60	Central refrigeration plant
		T61	Primary/Secondary cooling distribution
T7	Local cooling/Refrigeration	T70	Local cooling units
		T71	Cold rooms
		T72	Ice pads

U Ventilation/Air conditioning systems

U1	Ventilation/Fume extract	U10	General supply/extract
		U11	Toilet extract
		U12	Kitchen extract
		U13	Car parking extract
		U14	Smoke extract/Smoke control
		U15	Safety cabinet/Fume cupboard extract
		U16	Fume extract
		U17	Anaesthetic gas extract
U2	Industrial extract	U20	Dust collection
U3	Air conditioning – all air	U30	Low velocity air conditioning
		U31	VAV air conditioning

		U32	Dual-duct air conditioning
		U33	Multi-zone air conditioning
U4	Air conditioning – air/water	U40	Induction air conditioning
		U41	Fan-coil air conditioning
		U42	Terminal re-heat air conditioning
		U43	Terminal heat pump air conditioning
U5	Air conditioning – hybrid	U50	Hybrid system air conditioning
U6	Air conditioning – local	U60	Free standing air conditioning units
		U61	Window/Wall air conditioning units
U7	Other air systems	U70	Air curtains

V Electrical supply/power/lighting systems

V1	Generation/Supply/HV distribution	V10	Electricity generation plant
		V11	HV supply/distribution/public utility supply
		V12	LV supply/public utility supply
V2	General LV distribution/ lighting/power	V20	LV distribution
		V21	General lighting
		V22	General LV power
V3	Special types of supply/ distribution	V30	Extra low voltage supply
		V31	DC supply
		V32	Uninterrupted power supply
V4	Special lighting	V40	Emergency lighting
		V41	Street/Area/Flood lighting
		V42	Studio/Auditorium/Arena lighting
V5	Electric heating	V50	Electric underfloor heating
		V51	Local electric heating units
V9	General/Other electrical work	V90	General lighting and power (small scale)

W Communications/Security/Control systems

W1	Communications – speech/ audio	W10	Telecommunications
		W11	Staff paging/location
		W12	Public address/Sound amplification
		W13	Centralized dictation
W2	Communications – audio-visual	W20	Radio/TV/CCTV
		W21	Projection
		W22	Advertising display
		W23	Clocks
W3	Communications – data	W3	Data transmission
W4	Security	W40	Access control
		W41	Security detection and alarm
W5	Protection	W50	Fire detection and alarm
		W51	Earthing and bonding
		W52	Lightning protection
		W53	Electromagnetic screening

W6	Control	W60	Monitoring
		W61	Central control
		W62	Building automation

X Transport systems

X1	People/Goods	X10	Lifts
		X11	Escalators
		X12	Moving pavements
X2	Goods/Maintenance	X20	Hoists
		X21	Cranes
		X22	Travelling cradles
		X23	Goods distribution/Mechanized warehousing
X3	Documents	X30	Mechanical document conveying
		X31	Pneumatic document conveying
		X32	Automatic document filing and retrieval

Appendix B ACE Agreements 1995 A(2), B(2) and C(2) Appendix 1, work elements correlated to CAWS

ACE Engineering (Building) Services Work elements	Nearest CAWS Reference
Acoustical design and treatment	*(1)
Air compressors and compressed air services	S30
Air conditioning and mechanical ventilation services	U
Automatic blinds and shutters	L11, L12, *(3)
Bedpan washing and disposal equipment	N20–23 *(3)
Boilers and auxiliary plants	T10–14
Calorifiers	*(1)
Central dictation services	W13
Central vacuum cleaning installations	S50
Clock installations	W23
Cold water services	S
Combined heat and power installations	None
Conveyor installations and equipment	X23
Cooling water services	T21
Distribution mains for any services	included in R, S & T
Electric lighting and power installations	V2
Electric generation plant and systems	V10
Electric substations and switch gear	V11, V12
Electrical transmission services	V11, V12
Energy management systems	W6
Exhaust gas treatment and flues	T10, T11, T13
Fire detection and alarm systems	W50
Fire protection services	S6, S7
Flood lighting systems	V41
Food preparation, cooking, storage and serving equipment	N12 *(2)

Fuel gas distribution systems	S32, S33
Heating systems	T3, T4
Hot water services	S11, S12, S13
Incineration plant	R33
Information technology (IT) systems	W1, W3
Intruder detection and alarm systems	W4
Laundry equipment and services	N20–23 *(2)
Lifts, hoists, and escalators	X10, X20, X11
Medical gas and vacuum services	S34, S50
Pedestrian mover systems (travelators)	X12
Pneumatic tube conveyor systems	X31
Power-operated louvres	None
Public address, personnel location and call services	W12, W11
Radio and TV reception services	W20
Radio and TV transmission services	W20
Public health and plumbing services	N13, R1, R2, R3
Radiography, and similar medical investigation and treatment plant	N20–23 *(2)
Refrigeration and cold store installations	T60, T61, T70, T71
Refuse collection, compaction, incineration and disposal systems	R31, R32, R33
Security and access control systems	W40, W41
Steam and condensate return services	S51
Sterilizing equipment	N20–23, *(2)
Street lighting	V41(4)
Telephone installations and exchanges	W10
Thermal insulation applied to the engineering services systems	*(1)
Vibration control applied to the engineering services systems	*(1)
Water filtration and treatment systems	S20, S21
Window cleaning and other external access equipment	X22

Notes:
(1)Covered under other CAWS references where the application/treatment forms part of meeting the specified requirements for that element.
(2)Included under CAWS References for: 'N – Furniture/equipment'
(3)Included under CAWS References for: 'L – Windows/rooflights/screens/louvres'
(4)Street lighting is only listed in ACE 1995 Agreement C (2), for a design and construct contractor.

Appendix C Suggested duties for a consultant appointed by a D & B contractor

It is assumed the contractor has received a services brief in the employer's requirements and intends to use a designer contractor for design development.

Tender period

1. Receive and appraise employer's requirements (building services brief).
2. Confirm to the D & B contractor the adequacy of the brief, and/or raise any queries for clarification.
3. Submit a design risk appraisal to the D & B contractor.
4. Suggest alternative criteria/systems that would (a) reduce the contractor's exposure to risk and/or, (b) be commercially beneficial.
5. Set any building services design criteria that were left to the D & B contractor's choice.
6. Select any systems that were left to the D & B contractor to choose.
7. Prepare draft scheme:
   (a) calculate and evaluate building services loads;
   (b) prepare schematics;
   (c) produce outline scheme drawings for services installer to tender upon.
8. Prepare concurrent with (7), information for other consultants:
   (a) establish plant room sizes, horizontal and vertical distribution routes, together with any floor and ceiling services spatial zones;
   (b) give approximate live and dead load requirements to the structural engineer with respect to building services;
   (c) advise of building services penetrations through structure and building fabric;
   (d) agree insulation (thermal U) values.
9. Discuss and agree with design team and D & B contractor the requirements for building envelope air tightness.
10. Produce a building services specification for the installer to tender upon comprising:
    (a) the employer's requirements (Brief) as modified by consultant and design team;
    (b) material and workmanship standards;
    (c) schedules of plant, equipment and terminals for installing tenderers to complete with duties, capacities and numbers (of plant and equipment); Schedules should include a column for the tenderers to insert make, model and manufacturer;

(d) specification of BMS, HVAC controls, and motor control centres with associated standard references to Codes and Guides, etc., e.g. CIBSE, BSRIA, Building Energy Management System (BEMS);
(e) specify requirements for the commissioning of each separate system and service. This should describe facilities to be incorporated in the HVAC systems for air and water regulation, and the specification of system preparation, i.e. flushing, chemical cleaning and water treatment;
(f) specify requirements for operating and maintenance manuals and/or computerized system of planned preventive maintenance;
(g) specify requirements for end user familiarization, training and instruction on operating the systems in normal, standby and emergency modes.

11. Coordinate with the D & B contractor proposed building services preliminaries and preambles for inclusion in the consultant's specification.
12. Prepare building services budgets on a basis agreed with the D & B contractor.
13 With or on behalf of the D & B contractor conduct negotiations with public and private utilities suppliers for incoming services, supplies, and outflows, in the required locations and of the necessary pressure and size. Seek initial advice/programme on the earliest availability of the utilities.
14. If required within the scope of work agreed with the D & B contractor prepare a specification and sized design drawings for underslab and external drainage. Note: The D & B contractor may wish to include this as billed work in the substructure.
15. Agree with the D & B contractor and architect what fittings are to be provided by the installation tenderers, i.e. sanitary, lighting and kitchen, etc.
16. If requested give advice on suitable designer/installer contractors for the tender list.
17. Provide a suggested building services tender summary and breakdown (see Appendix F).
18. Evaluate designer/installer tender proposals and report to D & B contractor The report should also cover evaluation of any alternative proposals submitted by tenderers.
19. Produce as requested by the D & B contractor the following information for their submission documentation:
    (a) Technical appraisal of employer's requirements highlighting proposals where criteria and system selection was left to the D & B contractor;
    (b) alternative – write up the merits of alternative proposals;
    (c) provide details of energy consumption from basic up to the level of full life cycle cost;
    (d) provide CVs and experience;
    (e) join with D & B contractor's team in preparing for presentation/interview.
20. Attend interview/presentation.
21. Support D & B contractor in post tender interviews and negotiations.

Design phase (D & B contractor award to designer installer start on site)

1. Assist D & B contractor in documenting the order for the designer/installer.
2. Attend design development meetings.
3. Monitor subcontractor's design, including appropriate audits.
4. Monitor BS information needs of other design team members, e.g. architect, structural engineer

and ensure needs are met through the designer installer.

5. Receive and approve via the D & B contractor the designer installer's working drawings, shop drawings, wiring diagrams, etc.
6. Comment on the installer's programme.
7. Provide an inspection and meetings representation service throughout the installation, commissioning, system proving, fine tuning, and handover periods, to an agreed indicative schedule of attendance.
8. Receive and respond to all technical queries raised by tenderers and the appointed designer installer with respect to building services.

Installation phase

1. Implement the site representation, see item (7) of design phase.
2. Inspect quality of installations for (a) compliance with design, (b) materials and workmanship.
3. Receive and respond to all technical queries raised by installers and their specialist with respect to building services.
4. Comment on installer's monthly claim for interim certification.
5. Attend and report upon the specified offsite inspections.
6. Comment on designer installer's method statements.
7. Comment on appropriateness of designer installer's proposed compliance inspection sheets.

Commissioning to handover phase

1. Comment on designer installer's proposed method statements for HVAC systems preparation.
2. Witness the results of system preparation.
3. Approve designer installer's pre-commissioning method statements and checklists.
4. Approve designer installer's commissioning method statements, logic networks, programmes and checklists.
5. Witness the specified requirements for repeatable flow rates, functions and performance.
6. Receive and evaluate commissioning results.
7. Receive and comment upon draft operating and maintenance manuals, planned preventive maintenance (PPM) system, and record drawings.
8. Receive and comment upon designer installer's method statement and programme for instructing clients' staff/operatives on the operation of the systems and familiarization of the installations.
9. Agree with the D & B contractor a programme of final inspections, record defect and carry out 'defect clearance inspections'.
10. Carry out 12-month defects inspection.
11. Carry out final 'defect clearance' inspections.

Note: D & B contractors with building services managers (BSMs) and site engineers will wish to tailor the consultants duties to avoid expensive overlap. The BSRIA TN8/94 division of responsibilities pro formas will be of assistance. See also Tables 2.2 and 2.3, and Chapters 9 and 11 on Commissioning and Handover respectively.

Appendix D Building services design risk: a matrix for identifying potential pitfalls

Consider the following aspects and circle the appropriate risk score in the relevant project value column. (Scores marked with an asterisk assume that the project is designed and staffed by competent designers and contractors appropriate to the services content.) High risk is indicated by a total score over 16.

		Overall value of building project			
		<£2.5m	£2.5–10m	£10–20m	>£20m
Proportional value of services	<15%		2	4	6
	15–35%	2	4	0*	0*
	>35%	6	8	0*	0*
Type of building	Offices				
	Factory			1	1
	Retail			1	2
	Residential			1	2
	Hospital	1	1	2	4
	Other	1	1	2	4
Is there air conditioning?	Yes	6	8	4	2
	No				
What type of air conditioning?	VAV				
	VRV VRF	1	1		
	Chilled beams	2	2		
	Chilled ceilings	2	2	1	1
	Displacement vent	2	2	1	1
What are the temperature performance criteria?	±2°C				
	±1°C	1	1	1	1
	±0.5°C	2	2	2	2
What are the humidity performance criteria?	±20% rh				
	±10% rh	2			
	±5% rh	3	3	3	3
What are noise performance criteria?	45NR				
	40NR	1	1		
	35NR	2	2	2	2
	>30NR	4	4	4	4

		Overall value of building project			
		<£2.5m	£2.5–10m	£10–20m	>£20m
Are there any complex unusual services other than air conditioning?	Yes	1	3	2	1
Are there any innovative design features?	Yes	1	3	2	1
Does the scheme contain multiples of the same design?	Yes			1	4
Are fans or pumps speed controlled using electronic inverters?	Yes	1	1	2	2
Is there a BMS?	Yes	1	2	1	1
Is there adequate space allowances for plant, risers and voids?	No	1	2	1	1
Is there a requirement for builders work airtight shaft or room?	Yes	1	2	1	
Does the scheme contain low level fume exhausts?	Yes	1	2	1	
Is the specification clear on responsibilities for firestopping?	No	1	1		
Capability and resources of designer	Unknown	1	1		
	Suspect		1	2	4
Construction programme speed	Normal				
	Fast		1		
	Very fast	1	2	1	1
Is designer appointed on standard duties?	'Spec. & Drwgs'	1	2		
Completeness and quality of design tender information	Average				1
	Poor	1	1	2	3
How much design development by specialist subcontractor?	Extensive	1	2	1	1
	Normal		1		
Has subcontractor qualified system performance in his tender?	Yes		2	1	
Risk assessment	**Total**				

Appendix E Building services manager – job description

	Contractual route			
	Traditional	D & B	MC/CM	Project management
Reports to a line manager who may change as projects mutate from office to site based. In the discharge of his duties the building services manager (BSM) may also make direct contact with the client, services consultant, estimator, commercial manager, purchasing manager, project and planning managers and approving authorities and organizations as appropriate to the contractual route, viz:	X	X	X	X
Support for business development activities	X	X	X	X
Support and attendance on prequalification and presentations	X	X	X	X
Establishing and maintaining contact with the building services industry (consultants, research organizations, service contractors, manufacturers and other specialists)	X	X	X	X
Evaluation of incoming enquiries and advising line manager of all matters pertinent to the building				

	Contractual route			
	Traditional	D & B	MC/CM	Project management
services content, compliance requirements, programme time, and alternatives	X	X	X	X
Contributing to the establishment and maintenance of a vendor database for expertise in (a) consultants, (b) designer contractors (c) construct only services contractors, organizations	X	X	X	X
Building services cost advice for initial budgets, and procurement of specialist cost plan services		X	X	X
Advice on selection of building services consultants, and designer contractors.		X	X	X
Advice on selection of 'construction only' services contractors	X	X	X	X
Advice on building services procurement routes		X	X	X
Monitoring the progression of design and its required output to programme		X	X	X
Advice to planners on design, construction and commissioning programmes	X	X	X	X
Lead role in the preparation of building services consultants schedule of duties and division of responsibilities		X		X
Lead role in the preparation, issue, query handling and evaluation of enquiries and tenders to building services contractors and specialists	X	X	X	X

	Contractual route			
	Traditional	D & B	MC/CM	Project management
Contributing to design reviews		X	X	X
Ensuring designers are aware of and retain responsibility for services design		X		X
Advising line manager of risks implied in clients brief and design proposals for buildings services	X	X	X	X
Preparation of contribution to proposal document(s)	X	X	X	X
Attendance at the project tender settlement meeting, as required	X	X	X	X
Involvement in post-tender negotiations as required	X	X	X	X
Advice on building services project staffing	X	X	X	X
Monitoring building services contractors' performance for compliance with specified and contractual requirements	X	X	X	X
Contributing to the drafting of the project quality plan	X	X	X	X
Commenting on design teams project quality plan or system particularly with reference to building services		X		X
Commenting on the building services contractors quality plan or system	X	X	X	X
Advising on the scope and content, invitation to tender, and evaluation of tenders for maintenance contracts	X	X	X	X
Liaising with the estimator/ commercial manager to ensure that building services tenders are analysed and information is added to the cost planning databank		X	X	X

	Contractual route			
	Traditional	D & B	MC/CM	Project management
Establishing for each project where building services forms a significant element, an agreed division of responsibility for the coordination, inspection and monitoring of services during the construction period. The division of responsibility shall cover the design team, construction services resources as well as the building services manager and their staff assigned to the project	X	X	X	X
Responsible with training department for ensuring CPD for self and subordinate staff	X	X	X	X
Individual characteristics				
Multi-service experienced professional fully versed in the management of design, construction, commissioning and project handover procedures. Sound knowledge in the commercial optimization and cost planning of building services system selection		X		X
Ability to influence through verbal and written communication and establish good working relationships. Well developed skills in motivating others – leadership. Resilient nature, an ability to get things put right, a sense of humour	X	X	X	X

Appendix F Breakdown of tender – summary of headings and listing of subelements for customization

	£
Sanitary plumbing, rainwater and drainage	
Hot and cold water services	
Gas, compressed air, vacuum, medical gases, steam condense and other piped gases	
Fire services (piped)	
Electrical and ancillary services	
Heating, ventilation and air conditioning	
Utilities	
External services	
Lifts	
Escalators	
Free issue/fix only, e.g. hosereels	
Final connections, e.g. to process machinery	
Commissioning } Repeat for each main service grouping, i.e. M, E, P. F&L	
Record drawings }	
O & M manuals }	
Instructing clients/tenants staff }	
Provisional sums }	
Aftercare – maintenance contract }	
Total to tender form	

Notes: The form this takes may have already been determined in the employer's requirements. In the absence of any predetermined information a breakdown of tender should be produced that:

- enables easy accurate like for like comparison appropriate for the scale, technical complexity and geography of the project;
- when analysed will provide useful data for a computerized cost planning system.

The following listing is for guidance in producing customized tender breakdowns, and while extensive are not exhaustive.
For each separate trade enquiry (e.g. mechanical, plumbing, sprinklers, electrical services) the list of tender breakdown items must include items for commissioning and handover documentation.

Listing of subelements for customizing tender breakdowns

Sanitary plumbing, rainwater and drainage

Foul drainage below building to terminate 1 m beyond building line, or nearer manhole
Surface water drainage below building to terminate 1 m beyond building line, or nearer manhole
RW gutters and downpipes
RW outlets and downpipes
Supply and fix sanitary fitments (including sanitary towel macerators)
Soil waste and vent pipes to terminate at slab at lowest level
Branches to soil waste and vent pipes

Hot and cold water services

Cold water storage and rising main
Drinking water services (including service to DW fountains chilled water and auto vending machines)
Cold water down services
Treated water system
Local gas/electric hot water heaters
Water treatment (softening, deionized, demineralization, etc.)
Central domestic hot water services
High temperature HWS to kitchen – e.g. dishwasher
Chlorination

Gas, compressed air, vacuum, medical gases, steam

Gas to boilers
Gas to kitchens
Gas to process
Bottled gas 0_2, N_2 (local medical gases)
Centralized medical gases and distribution
Vacuum
Pneumatic document handling (e.g. Lamson tubes, etc.)
Compressed air
Centralized soap supply
Steam and condense

Fire services (piped)

Sprinklers
Hosereels
Dry fire riser
Wet riser
Auto CO_2
Halon
Emergency drenchers
Fire hydrant main
Foam inlets

Electrical and ancillary services

Substation
Transformers
HV Switchgear
LV Switchgear
Mains distribution
Sub mains distribution
Lighting (including luminaires)
LV power
Wiring to mechanical, public health and fire services
Emergency power – standby diesel generator
UPS
Emergency lighting
Fire alarms, manual/auto
Conduits, cable tray and circuit ways for voice/vision data
Conduits, cable tray and circuit ways for internal telephones
Wiring to lifts and hoists
Clocks
Staff call
Public address
Centralized aerial system for radio and TV
Heat/smoke detectors
Lighting fittings, tubes and lamps
Lightning protection
Snow melting
Ramp heating
Security services (CCTV, intruder alarms, doorphone, card access, etc.)
Earth leakage protection
Landlords metering
Tenants metering

Electric heating
Kitchen ventilation (hoods, wall/window fans)
Toilet/bathroom vent (local systems for each residential unit)

Heating ventilation and air conditioning (HVAC)

Elements (heating)
Boiler plant (alternative title, work in boiler room)
External chimney
Oil storage
Primary pumped circuit to HWS calorifiers/cylinders (alternative addition to title, including storage)
Low pressure hot water heating (outside boiler house) (including radiators, convectors, pipework to AHUs, etc.)

Systems (types associated with H&V)
Low pressure hot water heating (complete)
Direct fired warm air heating
Gas fired radiant heating
Toilet supply and extract ventilation
General ventilation (lift motor rooms, stores, etc. plant rooms, substations, switchrooms)
Car park ventilation
Kitchen ventilation (domestic/commercial)
Pressurization (stairs/escape routes)
Smoke management ventilation (atria, shopping centres, factories – may comprise both supply and extract systems)

Elements (cooling)
Refrigeration plant (including aircooled condensers)
Condenser water (pumps and pipework)
Cooling towers
Chilled water (pumps, tanks and distribution pipework)
Water treatment
Air handling units (AHUs)
Chilled beams
Chilled ceilings
Ice storage

Elements (general)
Ducting
Distribution pipework
Silencers
Insulation (ducting)
Insulation (pipework)

Terminals (fan coil units VAV units induction, chilled beams, grilles and diffusers)
Louvres (air intake and discharge)

Systems (types associated with air conditioning)
Air conditioning (complete systems including plant, distribution and terminals, e.g. fan coil, VRV, VAV, Versatemp, grilles and diffusers)
Displacement ventilation
Ventilation systems (including pressurization)
Smoke management

Controls
Controls
BMS/BEMS

Utilities

Connections to water main
Sprinkler main connection
Incoming gas main
Electric company charges (a) connection (b) contribution
Abandonment of existing services
Diversions of existing services
Facilities for incoming telephone and data services (containment and BWIC)

External services (including extending utility connections)

External water mains into building
External gas main into building
External lighting
Car park lighting
Hydrant main
Garden watering/irrigation
Lightning protection (see also electrical)

Lifts

Passenger
Goods (including scissors)
Disabled persons
Fireman's

Escalators

Façade maintenance
Motorized cradles and runways

Free issue/fix only

Allow for taking into storage/taking delivery, assembly, installation, testing and commissioning when installed of the following items of plant/equipment which will be provided as 'free issue':

Sanitary fittings
Lighting fittings
Kitchen equipment
Laundry equipment
Process machinery

Final connections

Kitchen equipment
Laundry equipment
Machine layout

Commissioning

System preparation – flushing and chemical cleaning
Water treatment
Commission and testing
Commissioning management

Spares

Record drawings – number of sets required ________

O & M manuals

Number of sets required
Computerized planned preventive maintenance ________

Instructing clients/tenants staff

Number of days

M _______
E _______
P _______
F _______

Provisional sums

First aid and hand fire appliances
special lighting in _______

Aftercare

12 months maintenance (direct contract with client)

Appendix G Declaration of management strategy requirements for a building services contract

Project ....

Contract for ....

Building services contractor ....

The objective in setting out these requirements is to inform of the manner in which we require your performance to be documented to us. The documentation required shall be evidence of good management, and progress in:

1. the preparation of quality, safety and environmental plans;
2. planning and programming the delivery of information;
3. the planning and programming of on- and offsite construction;
4. the procurement and production and approval of samples, mock-ups, trial site assemblies, etc.;
5. the control of work through supervision and inspection;
6. the confirmation of construction progress;
7. the management and progress of commissioning;
8. the management and recording of final inspections;
9. the planning and programming of the production of manuals and drawings for handover;
10. the planning and programming, and progress recording of instruction and training of end users in the operation and maintenance of the BS installations;
11. the setting up and management of any post-contract maintenance arrangements;
12. confirmation of the understanding of specified post contract responsibilities associated with fine tuning and system proving.

This is a framework setting out the scope and content of the documentation that will be necessary. It is not necessary to provide samples for approval at this stage, only a commitment to the proper documentation of the building services works. In principle we expect you to manage your works and present them to the specified standards. We require your cooperation to plan, organize, coordinate and control your works at the interfaces with other elements not forming part of your contract with us.

The following is an expansion of the framework outlining the type of documentation we require you to provide to us. The degree of detail to be agreed well in advance of its requirement to be used on site.

The outline information below is not exhaustive but indicative of the range of information required.

1.0 Preparation of quality, safety and environmental plans

Identify from the documents referred to in our order to you the requirements for the preparation, submission and approval of quality, safety and environmental plans. In the absence of any specified requirements your safety plan shall include details of your risk assessment and management procedures together with proposals for compliance with the Construction (Design & Management) Regulations 1994.

2.0 Planning and programming delivery of information

Comprising but not limited to:

2.1 Working drawings
2.2 BWIC drawings
2.3 Method statements
2.4 Approval of samples
2.5 Test and inspection plans (may be included with method statements)
2.6 Procurement schedules
2.7 Inspection and testing of offsite manufacture

3.0 Planning and programming of on- and offsite construction

Comprising programmes which, if necessary, must contain detailed activity sequences. Programmes showing first, second and final fix or just M & E by locational level are unlikely to be considered adequate for this project.

4.0 Programming the procurement, production and approval of samples, mock-ups and trial site assemblies, etc.

The BS contractor is required to summarize all specified requirements for the above and produce programmes showing dates including those for approval.

5.0 The control of work through supervision and inspection

The BS contractor shall produce an organogram for the on-site control of the work which shall be supported by work inspection checklists, defects logging and clearance procedures. You will be required to attend our site progress meetings at intervals to be agreed. We enclose a typical agenda for the 'first and second half' periods of the contract programme. In addition you are required to complete and return to us prior to each site meeting a contractor's report in the form of the samples enclosed which are again for first and second half contract periods.

6.0 The confirmation of construction progress

By the submission of progress status schedules, including records of construction pressure testing.

7.0 The management and progress of commissioning

You will be required to prepare a commissioning programme for the works of your contract which will be adjusted as necessary with your agreement, to integrate with the commissioning of other building services, and building works which are expected to be primarily finishing trades.

Where you are specified as having been assigned a commissioning management role you will be required to prepare coordinated programmes for all building services.

You will be required to chair/attend commissioning meetings which shall be on a regular basis, possibly as frequently as weekly.

In addition to the specified requirements of recording commissioning results and obtaining witness approvals, we require you to submit commissioning progress reports covering the status of system preparation (flushing, chemical cleaning and water treatment, etc.), system regulation, controls and BMS commissioning, witnessing and approving, etc.

8.0 The management and recording of final inspection

It is an essential requirement that you agree with us proposals for finally inspecting your own work, recording and clearing defects and providing progress status information.

9.0 The planning and programming of manuals and drawings for handover

You are required to plan backwards for the preparation of the operating and maintenance manual and record drawings, etc. and all related specified requirements. You shall allow in your programming for three four-week approval cycles of the documentation.

Within four weeks of your appointment you are required to submit to us a summary of the tender requirements for handover documentation.

Throughout the period of producing documentation we shall require from you status schedules showing the progress of preparation and approval.

10.0 The planning and production and progress recording of instruction and training of end users in the operation and maintenance of the installations

Within six weeks of our order to you we require a summarized abstract from the specified requirements covering the obligations to provide instruction and training. Six weeks prior to the implementation of any specified instruction and training you are required to submit your proposals, organization and programming for discharging these responsibilities.

At weekly intervals during the instruction and training programme we shall require a report/status schedule.

11.0 Setting up and management of any post-contract arrangements

Eight weeks after receipt of our order you are required to submit a statement confirming the specified requirements you are responsible for meeting. Eight weeks before handover you are required to remind us of any obligations you consider we may have under our contract with you for the discharge of these post-contract maintenance arrangements.

12.0 Confirmation of the understanding of specified post-contract responsibilities associated with fine tuning and system proving

Six weeks prior to handover please advise us of your specified involvement with the end user and designer in making fine tuning and system proving arrangements post contract.

Declaration of commitment

We agree to develop and implement a management strategy that meets the above requirements

PRINCIPAL CONTRACTOR	BS CONTRACTOR.....................
SIGNED BY	SIGNED BY............................
STATUS	STATUS ....
DATE.......................................	DATE ....

Appendix H Quality plans

N.G. BAILEY & CO LTD	QUALITY CONTROL PLAN	QUALITY PLAN NO. QP1 - E CONTRACT NO. G6-66024 SHEET 1 OF 12
CLIENT:- Wimpey Construction (UK) Ltd	CONTRACT:- Kingspool Development, York	SUBJECT:- Electrical Installation

INDEX

1.0 CONTRACT REVIEW
2.0 MATERIAL PROCUREMENT
3.0 MATERIAL AND EQUIPMENT RECEIPT
4.0 STORES CONTROL
5.0 ELECTRICAL EQUIPMENT INSTALLATION
6.0 SUPPORT STEELWORK INSTALLATION
7.0 CABLE INSTALLATION
8.0 EARTHING
9.0 TEST & COMMISSION
10.0 HANDOVER DOCUMENTATION

INSPECTION/TEST CODE

A1 - 100% INSPECTION OR TEST
A2 - SAMPLE INSPECTION OR TEST
W1 - 100% WITNESS OF AN INSPECTION OR TEST
W2 - SAMPLE WITNESS OF AN INSPECTION OR TEST
S - OPERATION SURVEILLANCE
H - HOLD POINT
V - 100% VISUAL INSPECTION
R1 - 100% DOCUMENTATION REVIEW
R2 - SAMPLE DOCUMENTATION REVIEW
X - SUBMIT DOCUMENTATION FOR RECORDS
D - SAMPLE DIMENSIONAL CHECK
N - NOTIFICATION POINT

REV	AMENDMENT DETAILS	DRAWN		DATE	CHECKED BY		DATE	APPROVED BY		DATE
		PRINT NAME	SIGNATURE		PRINT NAME	SIGNATURE		PRINT NAME	SIGNATURE	
0	For Approval	K. VARLEY	[signature]	11-10-93	I. MAY	[signature]	25.10.93			
A										
B										
C										
D										
E										

KV/FL/MANKING/D97

N.G. BAILEY & CO LTD	QUALITY CONTROL PLAN	QUALITY PLAN NO. QP1 - E CONTRACT NO. G6/66024 SHEET 2 OF 12
CLIENT:- Wimpey Construction	CONTRACT: Kingpool Development, York	SUBJECT:- Electrical Installation

ITEM NO.	OPERATION	OPERATING METHOD	ACCEPTANCE CRITERIA	INSP/CODE			VERIFYING DOCUMENT	REMARKS
				NGB	WCUK	PDG		
1.0	CONTRACT REVIEW							
1.1	Receive contract documentation	C/OC/014	Documents are as described on the transmittal note/ accompanying letter	R1	--	--	Transmittal note or accompanying letter	
1.2	Review contract documentation for any changes from the tender submission	C/OC/014	Requirements are in accordance with the tender submission	R1	--	--	Signed contract	
1.3	Prepare Quality Control Plan	C/OC/034	--------------------	---	--	--	-------------------	
1.4	Issue Quality Control Plan to controlled distribution	C/OC/034	--------------------	X	R1		Issue records	

KV/FL/MANKING/D97

N.G. BAILEY & CO LTD	QUALITY CONTROL PLAN	QUALITY PLAN NO. QP1 - E CONTRACT NO. G6-66024 SHEET 3 OF 12
CLIENT:- Wimpey Construction (UK) Ltd	CONTRACT: Kingpool Development, York	SUBJECT:- Electrical Installation

ITEM NO.	OPERATION	OPERATING METHOD	ACCEPTANCE CRITERIA	INSP/CODE			VERIFYING DOCUMENT	REMARKS
				NGB	WCUK	PDG		
2.0	**MATERIAL PROCUREMENT**							
2.1	Complete material requisition stating quantities, types and specification requirements	C/SP/016	Correctly completed material requisition	R1	--	--	Material requisition	
2.2	Submit material requisition for authorisation	C/SP/016	Authorised material requisition	H	--	--	Material requisition	
2.3	Submit authorised material requisition to Purchasing for procurement	C/SP/018	Purchase order raised	---	--	--	Purchase order	
2.4	Conduct works inspections as necessary	C/OC/036	Purchase order details	W2			Visit report	

KV/PL/MANKING/D97

N.G. BAILEY & CO LTD	QUALITY CONTROL PLAN	QUALITY PLAN NO. QP1 - E CONTRACT NO. G6-66024 SHEET 4 OF 12
CLIENT:- Wimpey Construction (UK) Ltd	CONTRACT: Kingspool Development, York	SUBJECT:- Electrical Installation

ITEM NO.	OPERATION	OPERATING METHOD	ACCEPTANCE CRITERIA	INSP/CODE			VERIFYING DOCUMENT	REMARKS
				NGB	WCUK	PDG		
3.0	**MATERIAL AND EQUIPMENT RECEIPT**							
3.1	Verify NGB purchased materials comply with the procurement details	C/OS/024	Compliance with NGB procurement details and zero damage	V,R1	--	--	Delivery Note Purchase Order	
3.2	Verify receipt of any necessary Certificates of conformity and Test Certificates	C/OS/024	Contract Specification Section 2 Part A Clause 1855	R1			Certificates of Conformity and Test Certificates	

KV/FL/MANKING/D97

N.G. BAILEY & CO LTD	QUALITY CONTROL PLAN	QUALITY PLAN NO. QP1 - E CONTRACT NO. G6-66024 SHEET 5 OF 12
CLIENT:- Wimpey Construction (UK) Ltd	CONTRACT: Kingspool Development, York	SUBJECT:- Electrical Installation

ITEM NO.	OPERATION	OPERATING METHOD	ACCEPTANCE CRITERIA	INSP/CODE			VERIFYING DOCUMENT	REMARKS
				NGB	WCUK	PDG		
4.0	**STORES CONTROL**							
4.1	Ensure materials and equipment are stored in accordance with the manufacturers instructions	C/OS/024	Manufacturers storage instructions	S	--	--	-------------------	Recorded by exception only

KV/FL/MANKING/D97

N.G. BAILEY & CO LTD	QUALITY CONTROL PLAN	QUALITY PLAN NO. QP1 - E CONTRACT NO. G6-66024 SHEET 6 OF 12
CLIENT:- Wimpey Construction (UK) Ltd	CONTRACT: Kingspool Development, York	SUBJECT:- Electrical Installation

ITEM NO.	OPERATION	OPERATING METHOD	ACCEPTANCE CRITERIA	INSP/CODE			VERIFYING DOCUMENT	REMARKS
				NGB	WCUK	PDG		
5.0	**ELECTRICAL EQUIPMENT INSTALLATIONS**							
5.1	Check equipment mounting and cable entry arrangements	--------	Contract Specification and Drawings	V	--	--	-------------------	Recorded by exception only
5.2	Position and install electrical equipment as detailed in the Contract Specification	C/CS/025	Contract Specification and	S	--	--	-------------------	
5.3	Inspect and test equipment in accordance with the Contract Specification	C/CS/039	Contract Specification and Drawings	A1	S		NGB Forms as applicable	

KV/FL/MANKING/D97

N.G. BAILEY & CO LTD	QUALITY CONTROL PLAN	QUALITY PLAN NO. QP1 - E CONTRACT NO. G6-66024 SHEET 7 OF 12
CLIENT:- Wimpey Construction (UK) Ltd	CONTRACT: Kingspool Development, York	SUBJECT:- Electrical Installation

ITEM NO.	OPERATION	OPERATING METHOD	ACCEPTANCE CRITERIA	INSP/CODE			VERIFYING DOCUMENT	REMARKS
				NGB	WCUK	PDG		
6.0	**SUPPORT STEELWORK INSTALLATIONS** Tray, Rack, Trunking, Conduit, etc.							
6.1	Check route for compliance with the Contract Specification and Drawings	---------	Contract Specification and Drawings	V	--	--	-------------------	Recorded by exception only
6.2	Erect cables support steelwork in accordance with the Contract Specification Drawings	C/OS/025	Contract Specification and Drawings	S	--	--	-------------------	

KV/FL/MANKING/D97

N.G. BAILEY & CO LTD	QUALITY CONTROL PLAN	QUALITY PLAN NO. QP1 -E CONTRACT NO. G6-66024 SHEET 8 OF 12
CLIENT:- Wimpey Construction (UK) Ltd	CONTRACT: Kingspool Development, York	SUBJECT:- Electrical Installation

ITEM NO.	OPERATION	OPERATING METHOD	ACCEPTANCE CRITERIA	INSP/CODE			VERIFYING DOCUMENT	REMARKS
				NGB	WCUK	PDG		
7.0	CABLE INSTALLATION							
7.1	Verify cable routes for compliance with the Contract Specification and Drawings	--------	Contract Specification and Drawings	V	--	--	------------------	Recorded by exception only
7.2	Install cables as detailed in the Contract Specification and on the Drawings	C/OS/025	Contract Specification and Drawings	S	--	--	------------------	
7.3	Inspect the cables in accordance with the Contract Specification	C/OS/039	Contract Specification	A1	--	--	------------------	Recorded by exception only
7.4	Gland and test the cables in accordance with the Contract Specification	C/OS/025 C/OS/039	Contract Specification	A1	--	--	NGB Forms as applicable	
7.5	Inspect glands and terminations in accordance with the Contract Specification and Drawings	C/OS/039	Contract Specification and Drawings	A1	--	--	------------------	Recorded by exception only

KV/FL/MANKING/D97

N.G. BAILEY & CO LTD	QUALITY CONTROL PLAN	QUALITY PLAN NO. QP1 - E CONTRACT NO. G6-66024 SHEET 9 OF 12
CLIENT:- Wimpey Construction (UK) Ltd	CONTRACT: Kingspool Development, York	SUBJECT:- Electrical Installation

ITEM NO.	OPERATION	OPERATING METHOD	ACCEPTANCE CRITERIA	INSP/CODE			VERIFYING DOCUMENT	REMARKS
				NGB	WCUK	PDG		
8.0	**EARTHING AND LIGHTING PROTECTION**							
8.1	Verify cable and tape routes and rod and bar positions for compliance with the Contract Specification and Drawings	---------	Contract Specification and Drawings	V	--	--	-------------------	Recorded by exception only
8.2	Install cables, tapes, rods or bars in accordance with the Contract Specification and Drawings	C/OS/025	Contract Specification Drawings	S	--	--	-------------------	Recorded by exception only
8.3	Inspect and test cables, tapes rods or bars in accordance with the Contract Specification	C/OS/039	Contract Specification Section 2 Part A Clause 1935	A1	--	--		
8.4	Inspect the cable and tape terminations in accordance with the Contract Specification and Drawings	C/OS/039	Contract Specification and Drawings Section 2 Part A Clause 1935	A1	--	--		

KV/FL/MANKING/D97

N.G. BAILEY & CO LTD	QUALITY CONTROL PLAN	QUALITY PLAN NO. QP1 - E CONTRACT NO. G6-66024 SHEET 10 OF 12
CLIENT:- Wimpey Construction (UK) Ltd	CONTRACT: Kingspool Development, York	SUBJECT:- Electrical Installation

ITEM NO.	OPERATION	OPERATING METHOD	ACCEPTANCE CRITERIA	INSP/CODE			VERIFYING DOCUMENT	REMARKS
				NCB	WCUK	PDG		
10.0	**HANDOVER DOCUMENTATION**							
9.1	Compile quality records in accordance with the Specification	C/OC/028	Contract Specification	R1	--	--	Document dossiers	
9.2	Submit quality records to the Client in accordance with the Contract Specification on the completion of site activities	C/OC/028	-------------------	X	R2		Document transmittal	

KV/PL/HANKING/D97

N.G. BAILEY & CO LTD	QUALITY CONTROL PLAN	QUALITY PLAN NO. QP1 - E CONTRACT NO. G6-66024 SHEET 11 OF 12
CLIENT:- Wimpey Construction (UK) Ltd	CONTRACT: Kingspool Development, York	SUBJECT:- Electrical Installation

ITEM NO.	OPERATION	OPERATING METHOD	ACCEPTANCE CRITERIA	INSP/CODE			VERIFYING DOCUMENT	DOC CODE	REMARKS
				NGB	WCUK	PDG			
10.1	Check and record continuity of ring final circuit conduits Method 1 Method 2		 IEE Regulations IEE Regulations	 A1 A1		 W2 W2	 NGB 182D NGB 183D	 C C	
10.2	Check and record continuity of Protective Conductors		IEE Regulations	A1		W2	NGB 184D	C	
10.3	Check and record Insulation Resistance and Polarity		IEE Regulations	A1		W2	NGB 185D	C	
10.4	Check and record the earth loop impedance		IEE Regulations	A1		W2	NGB 186D	C	
10.5	Check and record the operation of any RCD's		IEE Regulations	A1		W2	NGB 187D	C	
10.6	Commission emergency lighting		BS 5266	A1		W2	NGB 642A	C	
10.7	Commission fire alarm installation		BS 5839	A1		W2	NGB 643	C	
10.8	Test fire alarm installation		BS 5839	A1		W2	NGB 644	C	

KV/FL/MANKING/D97

N.G. BAILEY & CO LTD	QUALITY CONTROL PLAN	QUALITY PLAN NO. QP1 - E CONTRACT NO. G6-66024 SHEET 12 OF 12
CLIENT:- Wimpey Consturction (UK) Ltd	CONTRACT: Kingspool Develoment, York	SUBJECT:- Electrical Installation

ITEM NO.	OPERATION	OPERATING METHOD	ACCEPTANCE CRITERIA	INSP/CODE			VERIFYING DOCUMENT	DOC CODE	REMARKS
				NGB	WCUK	PDG			
9.9	Test and Record operation of Emergency Lighting		BS 5266	A1		W2	NGB 646	C	
9.10	Certify Testing and Commissioning of Electrical Installation		IEE Regulations	X		W2	Completion Certificate	C	

KV/FL/MANKING/D97

Suggested QP for unregistered BS contractors carrying out simple small works

It is suggested the following document could form the basis of a QP to be offered for adoption by building services contractors unregistered to BS 5750 Part 2, engaged to carry out simple trade works. It is assumed the contract value is within their trading range and they will not need to take on additional resources to carry out the work.

The adoption of a QP along these lines should not relieve the contractor of any of their contractual responsibilities.

Quality plan

Company: ....

Address: ....

....

....

Postcode: ....

Tel no: ....

Fax no: ....

Quality plan
For: .... (Trade)

At: .... (Site)

Prepared by: ....

Signed: ....

Date: ....

Issue No	Date	Revisions	Sign

Contents

1. General
2. Quality objectives
3. Responsibilities
4. Method statement work instructions
5. Inspection and testing records
6. Qualifications of operatives
7. Test equipment
8. Storage, protection and conformance of materials
9. Protection of our work
10. Sample materials, mock-ups
11. Offsite inspection/test
12. COSHH regulations
13. Hazardous operations
14. Quality control forms
15. Records

Appendix A – site organization

Quality plan

1. General
   This quality plan is for .... work to be carried out
   by .... for ....
   at ....
2. Quality objectives ....
   The work will be carried out to comply with the following:
   (a) contract specification/bill of quantities;
   (b) contract drawings as issued to us;
   (c) other instructions issued under the contract;
   (d) any relevant British Standard.
3. Responsibilities
   The person responsible for progress, workmanship and quality on the contract will
   be ....
   Position in company:..
   and any complaints beyond his authority will be dealt with
   by ....
   Position in company: ....
4. Method statement/work instructions
   The work will be carried out by our operatives in accordance with one or more of the following:
   (a) method statement;
   (b) work instructions where deemed necessary;

(c) the relevant British Standard;
(d) manufacturer's instructions.
When work is sublet or is of a complex nature, a method statement will be submitted to ....for approval prior to commencement.
The following are deemed to fall into this category.
(a)..
(b)..
(c)..

5. Inspection and testing records
Copies of the inspection sheets proposed to be used by us and our sub-contractors will be submitted to ....for information/approval. Similarly, where it is a requirement of ....or the specification to inspect and/or test any part of the work offsite, inspection test sheets shall be submitted to ....at least 7 days before the test/inspection is to take place. Signed inspection and testing records will be submitted to ....to confirm that the work has been carried out in accordance with the contract drawing/specification and relevant 'hold points' established.
6. Qualification of operatives
We confirm that all operatives employed by us have the necessary skills and training for the tasks to be carried out. Should any of the work be sublet to another contractor this will be agreed with ....prior to commencement and the above comments regarding skills will also apply.
7. Test equipment
Any test equipment used during the course of our contract will have been calibrated and tested within a time scale relevant to the equipment in use, or as required by the specification and manufacturer.
8. Storage, protection and conformance of materials
All material for incorporation in the works will be stored and protected in accordance with the specified requirements, good practice / manufacturer's recommendation / BS: 8000. Where required, a certificate of material conformance will be provided.
9. Protection of our own works
Protection of our own work will be carried out as required under the contract agreement with
....
10. Sample materials, mock-ups
In accordance with the specification a sample/mock-up of the following work/materials will be provided:..
11. Offsite inspection test
In accordance with the specification offsite inspection/tests will be provided for: ....
....
12. COSHH regulations
The use of any materials which are considered to be hazardous under the COSHH regulations will be advised in advance to ....together with any assessment sheets/control measures required.

13. Hazardous operations
    We have identified the following operations which we consider to be hazardous on this subcontract and for which proposed safe methods of working will be agreed with you: .....
    ....
14. Quality control forms (including inspection sheets)
    The attached/listed forms are used for quality control or where applicable ....-sheets will be used.
15. Records
    Quality records will be retained for a period of ....years.

Appendix A

Quality plan

Site organization

Contracts manager...

Site supervisor ....

Site foreman ....

Trades/labour...

Appendix I Programming – range of BS activities

These will vary according to building function, form and location

Item	1st Fix	2nd Fix	Final connections and terminals
EXTERNALS			
Meter pits at boundary Ducts/trenches to and between buildings(s) External drainage Utilities connections off site Ducts/trenches/containment – for external water, lighting, security features Earth rod pits	BWIC and installation of building services timing of activity will depend on: site layout, site roads, availability of utilities, extent of hard/soft landscaping Gas, water, electrics and drains must be on for wiring up mechanical services panels before system preparation		
ON BUILDING FACE			
Gutters and downpipes (including RWOs)	Continuous +	Connect to drain	
Lightning protection tapes including roof	*	*	Spikes, etc.
External lighting (safety, security, and feature)	*	*	*
CCTV cameras	*	*	*
Presence detectors	*	*	*
Sensors for HVAC	*	*	*
Louvres for air intake/discharge	←——— varies ———→		
Weatherings/flashings to roof penetrations	To secure earliest weathertight		
Telecommunication aerials/TV	←——— varies ———→		
INTERNALS (for watertightness)			
Soil, waste and vent pipe	*	Roof penetrations affect weathertightness	
Rainwater downpipes	*		
Dry fire riser, hosereel riser	*		
Sprinkler riser	*		

Item	1st Fix	2nd Fix	Final connections and terminals
PLANT			
Plant into lowest watertight levels, e.g. boilers, chillers, compressors, pumps, cylinders, tanks, generators, etc.	*		
Plant and equipment into 'guaranteed' watertight lowest levels, e.g. transformers, LV/HV switchgear, motor control centres, UPS	*		
Plant and equipment as above into watertight upper levels	*		
Plant and equipment as above into watertight roof level plant rooms	*		
MECHANICAL			
Prefabricate and install plant room piping	*		
Prefabricate and install sheet metal ducting	*		
NB including motorized valve bodies/ dampers	*		
Piping distribution from mechanical plant (Horizontal and vertical main runs)	*		
Ducting distribution from AHUs/fans (Horizontal and vertical main runs)	*		
Sprinkler mains distribution (Horizontal and vertical main runs)	*		
Pressure test all piping in sections	* Varies	*	
Pressure test all ducting in sections	*		
Insulate ducting		*	
Insulate piping		*	
Install HVAC terminals – rads convectors, VAV terminals, FCUs	*		
Pipe up HVAC terminals "		* varies	*
Duct up HVAC terminals "		* varies	*
ELECTRICAL			
Install main electrical distribution to distribution boards	*		
e.g. 1. tray, ladder support systems	*		
2. trunking and conduit containment	*		
3. cable (heavy-on cleats)	*		
4. twin and earth (without containment)		*	*

Item	1st Fix	2nd Fix	Final connections and terminals
Position and fix domestic scale electric heaters, HWS Heaters, fans, etc., install submain distribution to trunking and conduit	*		
Wire out the above		*	*
Cable and wire out (1) and (2)		*	
Install containment trunking and conduit for fire, security, PA, telecoms and data	*		
Wire out above by 'specialists'		*	*
Install mains distribution to mechanical services plant rooms and lift motor rooms	*		
Wire up MCCs in mechanical plant rooms			*
Install tray, trunking, conduit in lift rooms and to mechanical services power drives from MCCs	*		
Wire out above (may be mechanical services own S/C)		*	*
CONTROLS WIRING – may be specialists:			
Fix motors and actuators, linkages, etc. to valves and dampers		*	
Fix sensors	*		
Install containment trunking/conduit	*		
Wire out containment		*	
Wire up (final conns)			*
MAINLY 2ND FIX AND FINALS			
Fix thermostatic rad valve tops			*
CW to local HWS heaters		*	
Position local gas/electric HWS Heaters	*		
Pipe up to sanitary fittings from local water heaters		*	*
Run out soil, waste and vent branches	*	*	
Pressure test soil, waste and vent pipes and branches	*	*	
Position and fix sanitary fittings		*	*
Fix seats, plugs, chains, tap tops		*	*
Install hosereels, dry and wet riser landing valves and pipe up		*	*

Item	**1st Fix**	**2nd Fix**	**Final connections and terminals**
Install sprinkler heads and pipe up		*	*
Pressure test wet fire systems		*	*
Fix grilles and diffusers			*
Fix luminaire shells, gear trays		*	
Wire up lighting		*	*
Earthing and bonding		*	*
Floor outlets boxes cut into tiles and positioned		*	
Wire up power to floor boxes Wire up data to floor boxes Wire up telecoms to floor boxes } may be by other specialist		*	*
TELECOMS AND DATA			
Install plant and equipment into weathertight clean dust free dry and conditioned to spec plantrooms e.g. PABX switches, frame rooms	*		
Install main distribution tray and trunking	*		
Carry out main horizontal and vertical distribution	*	*	
Install closet panels and wire up		*	
Install secondary distribution tray and trunking	*	*	
Cable out above and wire to floor outlets		*	*
Test cabling			*
GENERATORS AND UPS (STATIC/ROTARY)			
These will interface with the essential services sections of electric switch panels. Generally the work will be by specialists in dedicated locations closely lagging the activities of the associated electrical works.			

Item	1st Fix	2nd Fix	Final connections and terminals
SPECIAL ANCILLARY ELECTRICAL SERVICES			
The following are usually wired out, terminated and tested by specialists, from containment (trunking/conduit) provided by the general electrical subcontractor:			
Public address	Generally		
Telephones			
Fire alarms	falling in		
Fire detection (heat/ smoke heads)	the category		
HVAC controls and building management (BMS)	of 2nd fix/final		
Security	connections		
Card access			
Door phones			
CCTV			
Intruder alarms			
TV			
LIFTS			
Plump shaft			
Install Guides			
Entrances			
Hydraulic cylinder			
Motor room equipment			
Trunking, wiring and trailing cables			
Lift car and roping			
Test			

Appendix J BS inspection forms

HADEN YOUNG		SITE INSPECTION CHECK LIST
PROJECT:		TM 17.1.23 Page No. DW 1.1
		System or element to be inspected: DUCTWORK
Date of Inspection:		

	ITEM, OBJECTIVE, STANDARD	REPORT, ACTION
1	Layout as drawings.	
2	Positions of supports and brackets are as the drawings and specification.	
3	Duct invert levels as drawings.	
4	Turning vanes fitted as drawings and with leading edge facing air flow.	
5	Duct branches fitted correctly and joints sealed. Opening in duct wall not smaller than branch size..	
6	Duct materials and stiffening complies with specification.	
7	Temporary open ends capped and no foreign matter in ducts.	
8	Brackets and supports comply with the drawings and specification.	
9	Supports to adequately support the ducts, particularly adjacent to dampers to prevent distortion.	
10	Supports must not foul joints or stiffeners.	
11	Drop rods to be vertical, ends cut back within the support angle or 10mm below nut.	
12	Ducting passing through walls is to be independently supported.	
13	Supports to allow full vapour barrier and must not obstruct the insulation.	
14	Vapour sealed ducts to have thermal break material between the support and the duct.	
15		

TM 17.1.23 Page No: DW1.1 Rev 1	Inspector's signature:	Date:

HADEN YOUNG		SITE INSPECTION CHECK LIST
PROJECT:	TM 17.1.23 Page No. DW 1.2	
	System or element to be inspected: DUCTWORK	
Date of Inspection:		

	ITEM, OBJECTIVE, STANDARD	REPORT, ACTION
1	In-line equipment installed as drawings.	
2	Heavy in-line equipment supported separately and adequately.	
3	Fire dampers correctly sealed into walls and floors.	
4	In-line equipment installed with correct flow direction.	
5	Access doors installed as drawings and specification.	
6	Check that all access doors provide effective access to adjacent equipment.	
7	Check free movement of dampers and adequate locking of handles.	
8	Check fastenings, pitch of rivets/spot welds/ self-tapping screws.	
9	Check flange joints are correctly bolted, cleated, gasketed as specification without buckling and that corners are sealed.	
10	Flanged joints do not occur within wall and floor penetrations.	
11	Check that non-setting mastic correctly applied to joints between ducts and the structure.	
12	Flexible duct connections to terminals are to be free of sharp bends and crushing. To be minimum practical length.	
13	Check that flexible ducts are the type specified.	
14	Check the finish of the ductwork and supports. Ducts to be free from dents, distortion and buckling.	
15	Check galvanising not peeling or burned and cut edges painted.	

TM 17.1.23 Page No: DW1.2 Rev 1	Inspector's signature:	Date:

HADEN YOUNG		SITE INSPECTION CHECK LIST
PROJECT:	TM 17.1.23 Page No. DW 1.3	
	System or element to be inspected: DUCTWORK	
Date of Inspection:		

	ITEM, OBJECTIVE, STANDARD	REPORT, ACTION
1	Check sections prepared for testing, particularly location of temporary sealing blanks.	
2	Check test carried out to specification, particularly test pressure, duration and compatability of test rig orifice size with leakage flow.	
3	Check calibration status of test gauges.	
4	Check that pressure test certificate details are correct, including extent of test. Check that all necessary parties have signed the test certificate.	
5		
6	Reinstatement / completion: All temporary test blanks removed and ducts re-installed.	
7	Check internal and external cleanliness of the ductwork.	
8	Check grilles/diffusers fitted correctly, particularly regarding squareness/flatness to wall/ceiling surface.	
9	Check linear diffusers are straight.	
10	Test holes drilled in correct positions and capped/plugged.	
11	Check all types of damper operate correctly. Demonstrate to Client / Authorities if required.	
12	Check earthing and bonding complete, where applicable.	
13	Check identification and labelling complete and to specification.	
14	Check all dampers left open ready for commissioning.	
15		

TM 17.1.23 Page No: DW1.3 Rev 1	Inspector's signature:	Date:

HADEN YOUNG		SITE INSPECTION CHECK LIST
PROJECT:	TM 17.1.23 Page No. IN 1.1	
	System or element to be inspected: INSULATION - GENERAL	
Date of Inspection:		

	ITEM, OBJECTIVE, STANDARD	REPORT, ACTION
1	Check materials comply with the specification and drawings.	
2	Check that the under-surface is properly prepared (wire brushing / painting / bonding)	
3	Check that the insulation is applied correctly and is of correct thickness.	
4	Check that the fixing / bonding to the under-surface is correct.	
5	Check that there are no gaps between sections.	
6	Check that valve boxes are adequetely packed with insulation.	
7	Check that the valve boxes are correctly installed without gaps and with adequate fastenings.	
8	Check that valve boxes allow access to handwheels.	
9	Check that finish / cladding is as drawings and specification and has adequate supports / fixings.	
10	Check that the cladding is free of dents, scratches, sharp edges and gaps.	
11	Check that the identification and labelling is as the specification.	
12		
13		
14		
15		

TM 17.1.23 Page No: IN1.1 Rev 1	Inspector's signature:	Date:

HADEN YOUNG		SITE INSPECTION CHECK LIST
PROJECT:		TM 17.1.23 Page No. PW 1.1
Date of Inspection:		System or element to be inspected: L.T.H.W. & CH. W. PIPEWORK

	ITEM, OBJECTIVE, STANDARD	REPORT, ACTION
1	Layout as drawings	
2	Positions of anchors/guides/supports/brackets are as the drawings/specification	
3	Correct levels/falls	
4	Alignment of pipework and in-line equipment	
5	Adequate vents and drains	
6	Welder IDs stamped/tagged adjacent to welds	
7	Cold-draw applied correctly, where applicable	
8	Sufficient space between adjacent services for expansion/contraction/thermal insulation	
9	Valves on in-line equipment to specification and installed as drawings. Check flow directions/access.	
10	Flushing facilities as drawing.	
11	Brackets/supports/anchors/guides of correct type and in correct positions	
12	Brackets and supports adjusted to support/restrain the pipework.	
13	Anchors/guides securely fixed and effective. Adequate clearance on guides.	
14		
15		

TM 17.1.23 Page No: PW1.1 Rev 1	Inspector's signature:	Date:

HADEN YOUNG		SITE INSPECTION CHECK LIST
PROJECT:		TM 17.1.23 Page No. PW 1.2
		System or element to be inspected: L.T.H.W. & CH. W. PIPEWORK
Date of Inspection:		

	ITEM, OBJECTIVE, STANDARD	REPORT, ACTION
1	Terminal equipment installed as drawings.	
2	Terminal equipment installed to manufacturer's recommendations.	
3	Terminal equipment connections in correct flow direction.	
4	Sufficient disconnect points (unions/flanges). Disconnect points on correct side of IVs.	
5	Nuts and bolts adequately tightened.	
6	Washers fitted under nuts and bolts.	
7	Correct bolt lengths.	
8	Welds painted.	
9	Screwed joints clean and protective treated (where applicable). No surplus joint materials.	
10	Protective coatings undamaged.	
11	Pipework painted to specification.	
12	No distortion of screwed fittings.	
13	Check pipework system for damage.	
14	For testing, check valves open and temporary stool pieces installed	
15	Tests carried out to specification, particularly test pressures and test durations. Check calibration status of pressure gauges.	

TM 17.1.23 Page No: PW1.2 Rev 1	Inspector's signature:	Date:

HADEN YOUNG		SITE INSPECTION CHECK LIST	
PROJECT:		TM 17.1.23 Page No. PW 1.3	
		System or element to be inspected: L.T.H.W. & CH.W. PIPEWORK	
Date of Inspection:			

	ITEM, OBJECTIVE, STANDARD	REPORT, ACTION
1	Check pressure test details are correct (including extent of test) and signed off by HY and the Client's Representative.	
2	Check that terminal equipment is fully protected prior to flushing or cleaning.	
3	Flushing/cleaning carried out to specification.	
4	Check Cleanliness Certificate supplied by Cleaning Sub-Contractor.	
5	Pipework re-instated after cleaning, temporary stool pieces removed, permanent items installed.	
6	Isolating/regulating valves operable and motorised control valves manually operable.	
7	Probes, pressure gauges, test points, etc. fitted correctly as drawings. Test points clear of insulation.	
8	Strainers clean.	
9	Earthing and bonding complete.	
10	Identification and labelling complete.	
11	Trace heating applied before insulation.	
12		
13		
14		
15		

TM 17.1.23 Page No: PW1.3 Rev 1	Inspector's signature:	Date:

HADEN YOUNG	SITE INSPECTION CHECK LIST
PROJECT:	TM 17.1.23 Page No. PW 3.1
	System or element to be inspected: GAS PIPEWORK
Date of Inspection:	

	ITEM, OBJECTIVE, STANDARD	REPORT, ACTION
1	Layout as drawings.	
2	Positions of supports/brackets as drawings/specification.	
3	Correct levels/falls.	
4	Alignment of pipework and in-line equipment.	
5	Drains fitted if required.	
6	Welder IDs stamped/tagged adjacent to welds.	
7	Valves on in-line equipment to specification and installed as drawings. Check flow directions and access.	
8	Brackets and supports of correct type and in correct positions.	
9	Brackets and supports adjusted to support/restrain the pipework.	
10	Terminal equipment installed as drawings.	
11	Terminal equipment installed as manufacturer's recommendations.	
12	Sufficient disconnect points (unions/flanges). Disconnect points on correct side of IVs.	
13	Nuts and bolts adequately tightened.	
14	Washers fitted under nuts and bolts.	
15	Correct bolt lengths.	

TM 17.1.23 Page No: PW3.1 Rev 1	Inspector's signature:	Date:

HADEN YOUNG		SITE INSPECTION CHECK LIST
PROJECT:	TM 17.1.23 Page No. PW 3.2	
	System or element to be inspected: GAS PIPEWORK	
Date of Inspection:		

	ITEM, OBJECTIVE, STANDARD	REPORT, ACTION
1	Welds painted.	
2	Screwed joints clean and protective treated (where applicable). No surplus joint material.	
3	Protective coatings undamaged.	
4	Pipework painted to specification.	
5	No distortion of screwed fittings.	
6	Check pipework system for damage.	
7	For testing, check valves open and temporary stool pieces installed.	
8	Tests carried out to specification and Gas Authority Codes of Practice, particularly test pressures and durations. Check calibration status of gauges.	
9	Check pressure test details are correct (including extent of test) and signed off by HY and the Client's Representative.	
10	Pipework re-instated after testing and temporary stool pieces removed	
11	Valves operable and motorised control valves manually operable.	
12	Probes pressure gauges, test points,etc. fitted correctly as drawings.	
13	Earthing and bonding complete.	
14	Identification and labelling complete.	
15	Ensure that the complete pipework installation is fully purged with nitrogen to remove all air before the admission of natural gas.	

TM 17.1.23 Page No: PW3.2 Rev 1	Inspector's signature:	Date:

	LIFT INSTALLATION CHECKS	SHEET 1 OF 2
SYSTEM / PLANT IDENTIFICATION		COMMENTS
1. Earthing of non-current carrying metal-work in machine room		
2. Phase fail and reversal relay		
3. Earth link or switch		
4. Push buttons on controller		
5. Rubber mats		
6. Framed and glazed notices		
7. Instructions		
8. Signs		
9. Journey recorders		
10. Machine room ventilation		
11. Machine room lighting		
12. Machine room access		
13. Lubricators		
14. Landing push buttons		
15. Indicators		
16. Maintenance telephone		
17. Operation of door or gate locks		
18. Operation of retiring cam		
19. Operation of doors or gates		
20. Push buttons and switches in car		
21. Alarm or emergency telephone		
22. Non-interference between controls		
23. Control system		
24. Earthing of non-carrying metal-work in car, well, and landings		
25. Escape hatch		
26. Ropes and crosshead label		
27. ON-OFF switch on car roof		
28. Slow switch on car roof		
29. Lamp socket on car roof		
30. Smoke vents		
31. Shaft separation		
32. Gate locks connected to positive		
33. Emergency release on gates or doors		
DISTRIBUTION:		INSPECTION DATE INSPECTORS NAME SIGNATURE

	LIFT INSTALLATION CHECKS	SHEET 2 OF 2
SYSTEM / PLANT IDENTIFICATION		COMMENTS
34. Safety gear clearances 35. Pit and equipment 36. Broken tape and rope switches 37. Basement door contact 38. Wiring and cabling 39. Painting 40. Finish (Car and Landings) 41. Special requirements		
DISTRIBUTION:		INSPECTION DATE INSPECTORS NAME SIGNATURE

Appendix K BS contractor reports – requirements

Use in conjunction with Table 8.1 Agenda for site meeting: from start on site to mid-construction.

Contractor's report no. .... Date ....

Name ....

This report is to be completed and returned to the site office no later than ....

You are requested to attend a contractor's meeting on ....
when this report will be discussed with your representative.

CONTRACTOR'S REPORT ON PROGRESS SINCE LAST MEETING

A reply is to be given to all questions. Support information may be attached providing it is clearly cross-referenced to the report section.

(1) Report on information
Since the last site progress meeting:

1.1 What quality, safety and environmental plans (with status) have you issued?

Title	Source	Date

1.2 What BWIC drawings have you issued 'for approval'?
Give details and dates of issue and approval required by.

Drawings/details	Issued	Approval required

1.3 What working drawings (including installation, coordination, wiring diagrams, shop and manufacturer's etc., etc.) have you issued? Give details, date of issue and approval required by.

Drawings/details	Issued	Approval required

1.4 What construction method statements (including safe working practice proposals) have you issued. Give details, source and date of issue.

Details	Source	Issue

1.5 What samples (including offsite visits) have been submitted? Give details and dates of issue and approval required by.

Details	Issued	Approval required

1.6 What test and inspection plans have been issued and offsite visits made?
Give details and dates of issue and approval.

Details	Issued	Approval status

1.7 What procurement schedules have you issued?

Details	Date

1.8 Are you delayed in any way, by lack of approvals?

(2) Information and/or decisions required from DT

2.1 What RFIs have you issued? Give details – dates of issue and information required by.

RFIs	Issued	Information required

2.2 What TQS have you issued? Give details and dates of issue and response required by.

TQS	Issued	Response required

2.3 Are you delayed in any way by lack of information?

(3) Progress of procurement and offsite production

3.1 What offsite visits for sample approvals, inspection and testing are due to be made before the next scheduled site progress meeting? Give details, firms expected to be represented and dates.

Details	Visiting	Attendees	Date

3.2 Are you able to obtain materials to comply with the current programme we have agreed with you?

3.3 What major deliveries have you scheduled in the period up to the next site progress meeting?

3.4 Do you require craneage? If so, state the sizes, weights and date required.

Items/materials	L × W × H × weight	Date

(4) Safety

4.1 Have you received any safety related notifications other than those issued by the PC? If so, give details, dates and attach a copy.

Details	Issued	Date received

(5) Site progress

5.1 What is your current programme position?

5.2 If you are in delay, substantiate.

5.3 List activities of work completed since the last report.

5.4 What is your labour strength currently employed on the contract?

5.5 How many operatives and supervisors will you require on site over the next four weeks to comply with the programe?

5.6 Outline any problems or delays you envisage.

(6) Inspection, construction testing and defect clearance
Summarize the change in status of the following since the last scheduled progress meeting.
6.1 Inspection.

6.2 Construction testing, e.g. pipework and ducting, pressure and air leakage; progressive electrical testing, etc., etc.

6.3 Defect clearance (including clearance of defects notified by PC, CoW, visiting Inspectors by regional utilities and LAs etc.).
(Note: Substantiation of statements may be called for.)

(7) Variations
7.1 What variation instructions have been received since the last scheduled site progress meeting. Give details and dates of issue and receipt.

Details	Issued	Received

BS contractor ....

Signed ....

Position .... Date ....

Use in conjunction with Table 8.2, agenda for site meeting: mid-construction to handover.

Contractor's report no. .... Date ....

Name ....

This report is to be completed and returned to the site office no later than ....
You are requested to attend a contractor's meeting on ....
when this report will be discussed with your representative.

CONTRACTORS REPORT ON PROGRESS SINCE LAST MEETING

A reply is to be given to all questions. Support information may be attached, providing it is clearly cross-referenced to the related report section.

(1) Safety

1.1 Have you received any safety related notifications other than those issued by the PC? If so, give details and dates and attach a copy.

Details	Issued	Date received

(2) Information and/or decisions required from DT

2.1 What RFIs have you issued? Give details and dates of issue and information required by.

RFIs	Issued	Information required

2.2 What TQS have you issued? Give details and dates of issue and response required by.

TQS	Issued	Response required

2.3 Are you delayed in any way by lack of information?

(3) Progress of procurement and offsite production

3.1 What offsite visits for sample approvals, inspection and testing are due to be made before the next scheduled site progress meeting? Give details, firms expected to be represented and dates.

Details	Visiting	Attendees	Dates

3.2 Are you able to obtain materials to comply with the current programme we have agreed with you.

3.3 What major deliveries have you scheduled in the period up to the next site progress meeting?

3.4 Do you require craneage? If so, state the sizes, weights and date required.

Items/materials	L × W × H × weight	Date

(4) Site progress

4.1 What is your current programme position?

4.2 If you are in delay, substantiate.

4.3 List activities of work completed since the last report.

4.4 What is your labour strength currently employed on the contract?

4.5 How many operatives and supervisors will you require on site over the next four weeks to comply with the programme?

4.6 Outline any problems or delays you envisage.

(5) Inspection, construction testing and defect clearance

Summarize the change in status of the following since the last scheduled progress meeting.

5.1 Inspection.

5.2 Construction testing, e.g. pipework and ducting, pressure and air leakage; progressive electrical testing, etc., etc.

5.3 Defect clearance (including clearance of defects notified by PC, CoW, visiting Inspectors by regional utilities and LAs etc.).

(6) Variations

6.1 What variations instructions have been received since the last scheduled site progress meeting. Give details and dates of issue and receipt.

Details Issued Received

(7) Commissioning

Note: For a small job of say up to nine months overall contract duration and basic services content, commissioning may be dealt with at the progress meeting. For jobs of longer periods and greater complexity for which separate commissioning meetings are arranged, it is only necessary to include a summary statement of commissioning progress here. What follows is the report format for a small job reporting on Commissioning in the general progress meeting.

7.1 Commissioning management

What commissioning management proposals, or changes to them, have been issued since the last site progress meeting? Give details and dates of logic diagrams, commissioning programmes and method statements issue and amendment.

Details Dates

7.2 What changes have been made to the management structure since the last site meeting to facilitate commissioning?

7.3 What *method statements* (including safe working practice, COSHH assessments and the like) have you issued since the last progress meeting? Give details, source, issue and ammendment dates for:

Details Dates

System preparation (eg blowing air, flushing & cleaning)

Pre-commissioning

Balancing (eg air and water flows)

Commissioning controls

7.4 What commissioning test and inspection plans have been issued and ammended, and offsite visits made, since the last site progress meeting? Give details and dates for:

Details Dates

Witnessing by client/DT

External Inspectorate (eg regional utilities,
L.A. inspector, insurers etc.)

7.5 Is commissioning being managed to the agreed programme?

7.6 Outline any anticipated problems or delays yoy envisage.

Substanciation of statements may be called for.

(8) Documentation (Manuals & Record Drawings)
As for commissioning, what follows is for a small job where progress on documentation is taken in the general site meeting.

8.1 What programmes and ammendments have been issued for the production of the specified documentation since the last site progress meeting. Give details and dates.

Details	Issued	Received

8.2 What samples for approval have been issued since the last scheduled site progress meeting. Gives details and dates of issue and approval for:

Details	Issued	Approval Required
Manual covers (material, lettering and binding)		
Manual Format (volumes, books dividers, indexing)		
Record drawing title block		
Wall hung charts, framed, glazed or plastic encapsulated		

8.3 Report on progress of production and approval of documentation. Give details and dates.

Details	Issued	Date & approval status

8.4 Outline any anticipated problems or delays envisaged.
(Note: Under the CDM Regulations the PC cannot discharge his responsibilities to the PS or the PS to the client end user until the H & S file is complete. The Contractor should be aware that it is a legal requirement for the H & S file to contain the operating and maintenance manuals and record drawings. The Contractor shall be held responsible for the consequential costs of non-compliance in the event of the PC being unable to hand over the H & S file on contract completion due to the unavailability of the BS documentation.)

(9) Training and instruction programme

9.1 What programmes and amendments have been issued for the specified training and instruction of client/end user personnel? Give details and dates of issue and approval for:

Details	Issued	Approval required

9.2 Report on change in status of delivery of training and instruction. Give details dates and reasons.

Details	Reasons	Dates

9.3 Outline any anticipated problems and delays envisaged.

(10) Final inspection and clearance of defects

10.1 What changes and amendments have been made to your proposals (method, programme and resourcing) for the final inspection and clearance of defects by your own supervision and inspection. Give details of changes since last report.

10.2 What is the status of progress? Provide summarized quantification of defects and assessment of percentage clearance.

(11) Handover

11.1 What is the status of specified requirement for handover of keys, tools and spares, lamps, tubes, fuses, etc.?

BS contractor ....

Signed ....

Position ....

Dated ....

Appendix L Commissioning management

BSRIA

4: SPECIFYING SYSTEM COMMISSIONING ACTIVITIES

Design Activity	Allocation of Responsibility		
	Designer	Installer	Other
Design			
4.1 Ensure that the selected systems will meet the client's brief and that their commissioning requirements are compatible with any project restraint concerning sectional handover/phasing.			
4.2 Identify and incorporate into system designs the essential components and features necessary to enable the proper preparation and commissioning of building services.			
4.3 Review all designs to ensure that systems can be properly prepared, and are commissionable.			
4.4 Prepare the ***commissioning specification.***			
Management			
4.5 Produce a commissioning method statement and logic diagram for integration into the building contractor's construction and finishes programmes.			
4.6 Produce a flushing, chemical cleaning and water treatment method statement, logic diagram and programme for integration into the building contractor's construction, commissioning and finishes programmes.			
4.7 Attend commissioning meetings as necessary OR Arrange and chair commissioning meetings as necessary.			
4.8 Comment on the adequacy of systems for commissioning as detailed on ***specialists' drawings*** and manufacturers' ***shop drawings*** prior to actual manufacture at works. Ensure comments are incorporated into finished products.			
4.9 Carry out site inspections, to ensure that the commissioning facilities are being installed. Check compliance with specified guides and standareds.			

BSRIA

4: SPECIFYING SYSTEM COMMISSIONING ACTIVITIES (cont)

	Design Activity	Allocation of Responsibility		
		Designer	Installer	Other
4.10	Monitor the on-going progress of the procurement, manufacture, installation and commissioning of all plant items.			
4.11	Assess the effects of any anticipated delays to the services installation and the completion of interfaces with the building works critical to the commissioning programme. Formulate strategies to overcome potential delays.			
4.12	Establish an agreed set of pro forma documentation relating to the commissioning and testing of plant and systems.			
4.13	Approve the proposed set of instruments for the commissioning and testing works.			
4.14	Ensure that the instrumentation is periodically calibrated as necessary and records retained.			
4.15	Witness the flushing, cleaning and treatment of systems in accordance with the ***commissioning specification.***			
4.16	Witness pre-commissioning activities in accordance with the ***commissioning specification.***			
4.17	a) Commission all systems to method, logic and programme (see 4.5) and record results.			
	b) Witness specified demonstration of system commissioning results.			
4.18	Witness and record the specified demonstration and testing of plant items and systems in accordance with the ***commissioning specification.***			
4.19	Establish with the building contractor procedures to enable the demonstration of normal emergency, shut down and standby mode operation of plant and systems.			

4: SPECIFYING SYSTEM COMMISSIONING ACTIVITIES (cont)

Design Activity	Allocation of Responsibility		
	Designer	Installer	Other
4.20 Witness demonstration of same to specified requirements.			
4.21 Witness the partial load testing of plant to the client and designer in accordance with the ***commissioning specification.***			
4.22 Witness the operation of the BMS on site to the specified requirements.			
4.23 Witness the functional testing of all safety interlocks in accordance with the ***commissioning specification.***			
4.24 Witness the demonstration of acoustic tests in accordance with the ***commissioning specification.***			
4.25 Witness the operation of plant and systems for specified periods of time to prove plant reliability.			
4.26 Produce commissioning report detailing the results of the commissioning and commenting on the performance of systems.			
4.27 Ensure that all plant settings are recorded, including appropriate reference to plant items. The records should be incorporated within the ***operating and maintenance manuals.***			
4.28 Accept completed systems.			

Appendix M Professional institutions, consultancy associations and learned societies

Name	Telephone	Range of information services												
		1	2	3	4	5	6	7	8	9	10	11	12	13
Association of Consulting Engineers	0171 222 6557	•1				•2			•	•	•			•2
Association of Consulting Scientists	0125 586 2412								•					
Association of Security Consultants	0181 998 6143	•	•	•	•	•	•		•					•
British Computer Society	0179 417 417		•				•		•	•	C	•		
British Institute of Facilities Management	01799 513371	•	•	•			•			•	•			
Chartered Institute of Building	01344 23355								•	•	B	•2		3
Chartered Institution of Building Services Engineers	0181 675 5211			•	•	•				•	C			
Health and Safety Executive	0114 289 2345	•	•	•	•9	•9	•9	•			•4			
Institute of Sport and Recreation Management	01664 65531	•	•	•	•	•	•		•	•	C			•
Institute of Energy	0171 580 7124										C			
Institute of Hospital Engineering	01705 823186	•	•		•	•				•	B	•		
Institute of Plumbing	01708 472791			•	•	•			•	•		•		•2
Institute of Quality Assurance	0171 730 7154			•				•		•	•			•
Institute of Refrigeration	0181 647 7033		•	•		•				•5				
Institute of Sound and Vibration Research	01703 592162													•
Institution of Electrical Engineers	0171 240 1871	•	•	•			•	•		•	•	•	•	•
Institution of Electronics and Electrical Incorporated Engineers	0171 836 3357									•	C			•
Institution of Fire Engineers	01162 553654							•6	•7	•				•
Institution of Gas Engineers	0171 636 6603	•	•	•	•	•	•			•		•		
Institution of Lighting Engineers	01788 576492		•			•				•		•		•
Institution of Mechanical Engineers	0171 222 7899							•		•	C	•	•	

Name	Telephone	Range of information services												
		1	2	3	4	5	6	7	8	9	10	11	12	13
Institution of Occupational Safety and Health	01162 571399									•	C	•	•	•8
Institution of Structural Engineers	0171 235 4535			•	•10				•	•		•2	•2	
Royal Institute of British Architects	0171 580 5533							•	•	•	•	•	•11	
Royal Institute of Chartered Quantity Surveyors	0171 222 7000			•				•	•	•	•	•	•	•
Water Management Society	01827 289558		•	•	•	•				•	C			

Range of information services

1. Standards
2. Codes
3. Guides
4. Technical memoranda
5. Technical notes
6. Bulletins/digests/newsletters
7. Bibliographies
8. Directory of members
9. Journal
10. Bookshop (B) Catalogue (C)
11. Library
12. Electronic database
13. Technical advice/centre

Notes to entries

1. Standards of professional practice
2. For members
3. Subscription information service construction information file
4. Bookshop is HMSO
5. Proceedings
6. Index of *Fire Engineer Journal* articles
7. Professional register
8. Services are primarily for members, although some technical enquiries from non-members are responded to
9. Incident and contract research reports
10. Reports
11. Bibliographic

Other information and data sources

The following organizations also provide information services covering variously, legislation, standards, codes and guides, products and research projects in the form of electronic databases available on subscription

Barbour Index	01344 884121
Oakland Consultancy (experts for industry, current research in academia)	01223 300475
ODI	0113 230000
Technical indexes	01344 426311
Other references	
British Library – business and information services	0171 3237454
The Met Office	01344 856836

Index

Page number appearing in **bold** refer to figures and page numbers appearing in *italic* refer to tables